Men Who Made a New Physics

Men Who Made

a New Physics

PHYSICISTS AND THE QUANTUM THEORY

Barbara Lovett Cline

With a new Foreword by Silvan S. Schweber

THE UNIVERSITY OF CHICAGO PRESS

Chicago and London

ACKNOWLEDGMENT is made to copyright holders for permission to use the following copyrighted material:

Excerpts and the diagram on page 239 from *Albert Einstein, Philosopher–Scientist,* Volume I. edited by Paul Arthur Schilpp. copyright 1949, 1951 by The Library of Living Philosophers. now published by The Open Court Publishing Company. La Salle. Illinois.

Excerpts from *Rutherford* by A. S. Eve. Cambridge University Press. New York, 1939.

Excerpts from the *Scientific American* in Chapter 13.

The University of Chicago Press, Chicago, 60637
The University of Chicago Press, Ltd., London

96 95 94 93 92 91 90   5 4

Library of Congress Cataloging in Publication Data

Cline, Barbara Lovett.
   Men who made a new physics.

   Reprint. Originally published: The questioners.
New York : T.Y. Crowell, 1965.
   Bibliography: p.
   Includes index.
   1. Physicists—Biography.   2. Physics—History.
3. Quantum theory—History.   I. Title.
QC15.C4   1987   530'.092'2   [B]   87–10786
ISBN 0–226–11027–3 (pbk.)

THE ADVANCES of the past thirty years in our understanding of the composition of matter at the subnuclear level constitute one of the great intellectual achievements of mankind. Physicists have formulated a quantum theory of the strong, electromagnetic, and weak interactions that accounts for all the data for energies below one 1 Tev. Only the unravelling of atomic structure in the period from the mid 1880s to the late 1920s, and more particularly the formulation of quantum mechanics in the period from 1925 to 1927, constitute a comparable achievement. I would argue that the latter was in fact a greater achievement and certainly a more revolutionary one. Greater because quantum mechanics is the foundation upon which all subsequent developments are based, and more revolutionary because it constituted a radical alteration in our conceptualization of the world, of how as "interveners" we interact with it and what as "representers" we know of it.

In this book Barbara Lovett Cline tells the story of how quantum mechanics was constructed and how its interpretation was put forward. The formulation of quantum mechanics was the work of a handful of people: Heisenberg, Dirac, Pauli, Born, Jordan, Schrodinger, Bohr. Never have so few achieved so much in so short a period of time. Their success was the culmination of three decades of intense research during which the domain of inquiry of physics became defined: the physicist's task was to elucidate the fundamental constituents of matter.

In the 1880s the problems of the constitution of atoms and molecules were addressed by both chemists and physicists. Two decades later, as the first two meetings of the Solvay Congress in 1911 and 1913 made evident, the problems connected with the structure of atoms were the physicist's province. How did this happen? Why were the physicists able to wrest the problems of the structure of matter from the chemists? Although Cline does not ask the question explicitly, it is one of the merits of her book that a partial answer to the query can be obtained from it.

By the end of the nineteenth century it had become apparent that both "chemical" atomism, which had helped elucidate the three-dimensional structure of molecules, as well as the "physical" atomism of the kinetic theory which had explained some of the thermal properties of gases, had encountered intractable problems. The leading physical chemists—Ostwald, Nernst, Arrhenius, Duhem—encouraged by the success of thermodynamics in explaining many features of chemical reactions without having to make any assumptions about the microscopic constitution of matter, became anti-atomists. Thus the subdiscipline of chemistry most able to address the problems of chemical structure did not do so because of these "ideological" commitments. On the other hand, the leading physicists—Boltzmann, Lorentz, and Planck—never wavered in their commitment to atomism. Additionally the chemists' philosophical predilection toward synthesis, as contrasted to that of the physicists' toward analysis, also probably played a role. Other, more internal, factors were also relevant. Toward the end of the nineteenth century both physicists and chemists were concerned with electrical conduction: the chemists with electric conductivity in solutions, and the physicists with electric conductivity in gases. However the conductivity of electrolytes in solutions is a much more complex problem than electrical conductivity in gases or solids. Only in the case of gases can ionization be readily studied. Because electrodynamic phenomena in general, and more particularly the phenomena connected with electrical conduction in gases (such as cathode rays) were the concern of physicists, physicists developed the apparatus and the technical skills (such as the ability to create high vacua) to address these problems. And in the process they discovered the electron and the nucleus, characterized their properties, and put forward the first models to account for the structure of atoms. Their control of the apparatus was perhaps the most important determinant in the control of the field. Part of the greatness of H. Hertz, J. J. Thomson, and E. Rutherford was their ability to design and construct the instruments and apparatus necessary to grasp the phenomena. It is apparatus and instruments that permit us to manipulate the unseen, micro-scopic "fundamental particles" that make up atoms and nuclei, and thus allow us to assert their reality.

It was the skill of the leading experimenters (e.g., J. J. Thom-son, E. Rutherford, J. Frank) as manipulator of phenomena, and the charisma of the physicists who headed the successful research groups at the leading scientific institutions—Rutherford at the Cavendish, Born and Frank in Göttingen, Sommerfeld in Munich, Bohr in Copenhagen—that decided the issue. These men trained most of the physicists who unravelled the structure of atoms and nuclei in the period from 1910 to 1940.

Cline's account is a sensitive and insightful one. The great advances, she tells us, have been as much experimental as theoretical ones. The apparatus that Thomson, Rutherford, Frank, and others designed were just as necessary for allowing us to see the world in novel ways as Schrodinger's, Heisenberg's, Dirac's, and Bohr's theoretical contributions.

The social dimension of the scientific enterprise comes through clearly in Cline's account. By adumbrating science as a way of life for Rutherford, the moral dimension of science is conveyed. By highlighting the philosophical commitments of the leading theorists, the metaphysical presuppositions that animate scien-tific activities are illustrated. By telling the story of the Caven-dish and of the Göttingen Institutes, the role that institutions play in shaping the way we come to interrogate nature is indicated. It is the passion, the temperament, the style of the leading contributors that Cline delineates best. The portraits of Einstein, Bohr, Pauli, Rutherford, Heisenberg, and Dirac that she sketches help us appreciate their genius. But because she concentrates almost exclusively on the "major figures," the picture of the social world of physics which is conveyed is somewhat distorted. *Men Who Made a New Physics* was clearly influenced by the view, prevalent among physicists, that great individuals are responsible for the great advances of the discipline. But even if one should disagree with that position, there is no doubt that the handful of physicists who created quantum mechanics were "off-scale" individuals, whose intellectual productions were singular; yet whose accomplishments are testimony to the capabilities of the human mind. Cline's presentation should make clear why the search for the fundamental constituents of matter—the physicist's attempt to find simplicity in the complexity the world presents—represents one of the most precious elements of our culture, and why the enterprise represents one of the few things that gives enduring meaning to our seemingly absurd lives.

Cline's stimulating introductory exposition of how quantum mechanics was created will make the reader want to explore in greater depth both the history and the philosophy of quantum mechanics. Surely that is an indication of how well she has succeeded.

S. S. SCHWEBER
February 1987

CONTENTS

ILLUSTRATIONS

In his novel *The Search*, C. P. Snow describes the reaction of a student in class on hearing the physics professor say he is not sure whether some of the subject matter in the course is right. This indication of disagreement among those inside physics comes as a surprise to the student; he has heard of past scientific controversies, but the current science which he is studying had seemed to lack them altogether, as if scientist-authorities backed it by unanimous vote. "Science," writes Snow, ". . . had seemed to be without people or contradictions."

The knowledge that physics is not as unanimous or bloodless as it may appear from the outside came as a surprise to me also. I wanted to know more—more, for instance, about the intellectual clash between Albert Einstein and Niels Bohr over a theory to which both contributed, the quantum theory.

A search began and this book represents what I learned. Specifically, it is about certain individuals who were working in physics with great success during the first quarter of this century when the quantum theory was shaped, as well as the theory of relativity. I have organized the book around the evolution of the first theory (to a lesser extent, the second), focusing on particular individuals, including Einstein and Bohr, who made contributions to the quantum theory, telling where and how they worked, the sort of people they were, and the different meanings which physics had for them.

An organization of this sort, where the general course taken by a science is seen through the contributions and careers of a few individuals, tends to convey the impression that science is the work of a few. That is a distortion which I have tried to minimize but was not able to eliminate; when one focuses on a few, the others will show up in the picture with less distinction, if they are caught within the frame at all.

I would like to acknowledge some of the help I received—from Philipp Frank's *Einstein: His Life and Times;* from Victor F. Weisskopf's *Knowledge and Wonder;* and from articles and other books which are listed in the back of this one. People acquainted with various aspects of the subject matter also helped me directly; they answered, elucidated; in some cases they read and criticized what I wrote. Without implying that they are responsible for any errors I might have made or that they share my biases I would like to thank Mrs. S. Hellmann, Mrs. C. John, H. Pauli, Mrs. B. Schultz, A. Bohr, G. Gamow, S. Goudsmit, D. Greenberg, J. Heilbron, G. Hevesy, M. Klein, O. Klein, C. Møller, R. Oppenheimer, D. Price, L. Rosenfeld, M. Shamos, and V. Weisskopf.

P. Hein kindly gave me permission to reproduce some of his drawings from the *Journal of Jocular Physics;* W. Heisenberg let me use some photographs. People at the Institute for Theoretical Physics (Copenhagen) and the Niels Bohr Library of the History of Physics and *Physics Today,* both at the American Institute of Physics, were helpful in a number of ways.

Last I would like to say thank you to R. Palter for early encouragement, to A. Petersen for opening my eyes to many things, and to C. Cline for advice and tact.

B. L. C.

Men Who Made a New Physics

CHAPTER ONE

Ernest Rutherford: Discovery

of the Nucleus

I've just been reading some of my early papers and, you know, when I'd finished, I said to myself, 'Rutherford, my boy, you used to be a damned clever fellow.'
—Lord Rutherford to Sir Henry Tizard

ONCE UPON A TIME, in Manchester, England, two physicists were talking about a student, Ernest Marsden, who was nineteen years old. "For some time now young Marsden has been helping me out in the laboratory," said the first physicist. "Don't you think he's ready for a small research of his own?"

The second physicist and director of the laboratory in question was Ernest Rutherford, an exuberant man with a big walrus mustache, a big frame, and a big voice. He knew "young Marsden" from the classroom—Rutherford was teaching physics at the University of Manchester then—as well as the laboratory.

Once he had accused the young man of disrupting an important experiment. For ten years, on and off, Rutherford had been trying to establish the identity of particles which are shot off from certain radioactive elements and which he had named "alpha." Now he was close to his goal as, with a spectroscope, he studied the spectrum produced by a sample of alpha particles collected in an evacuated glass tube, and recorded it in photographs. One day while this experiment was in progress Ruther-

1

ford, upon entering his workroom and going to his spectroscope, noticed that the instrument's prism was out of place. In another part of the room, Marsden was working on something at the optical bench. No one else was about.

Suddenly, Marsden felt a hand on the back of his neck. It was placed, he said, "only moderately gently." Then he heard the familiar big voice as Rutherford, in a fury, thunderously accused: "Did you move that prism?"

"No," Marsden replied calmly. From past experience he knew that there was nothing to fear and he rather enjoyed seeing the boss in a tantrum.

Sure enough, within half an hour Rutherford, who had left the room to find the guilty party, returned and sat down very deliberately next to Marsden, to ask his pardon. Then they had started to talk about their work and as Marsden listened to Rutherford's ideas and Rutherford listened, just as hard, to what he had to say, the young man had forgotten who was boss, who student.

On the day when Rutherford's assistant Hans Geiger asked him about "a small research" for Marsden, Rutherford replied that he had been thinking about the same thing. Then he suggested an experiment. Patience and care were required, not to mention good eyesight; but Marsden, he thought, could handle it. The young man was to observe the deflections of alpha particles sent at a target of matter.

At this time, 1909, the electron was the only known atomic particle. In an attempt to explain properties of matter, physicists had thought out a number of possible arrangements for electrons inside the atom. These arrangements, or atomic models, had to account for the fact that matter, generally speaking, is electrically neutral and therefore, to counteract the negatively charged electron, the atom must contain positive electricity in some form. A positively charged particle comparable to the electron was not known to exist; it was possible that positive electricity took a different form; perhaps it was a fluid. J. J. Thomson, an English physicist who was among the first to identify the electron, had worked out an atomic model based on this idea. His atom contained a fluid of positive electrification. Within this were embedded a sufficient number of electrons to offset the positive electricity. Thomson's atomic model appeared reasonable and was based on experimental findings. But this evidence was not plentiful. Much remained to be shown and Rutherford, with his alpha particles, had the best way then available to show it.

Alpha particles are smaller than atoms, but heavy (massive); and they are shot off from radioactive substances at great speeds. Therefore they could serve as high-energy bullets for probing atoms at a time when such bullets could not be produced by artificial means. As well as trying to identify his alphas, Rutherford was using them as experimental tools. In his laboratory, beams of alpha particles were sent through atoms to hit a detecting screen, so that one could observe whether or not the particles had been knocked off course during their intra-atomic journey.

This method of exploration has been likened to shooting bullets through a bale of hay in which a very small piece of platinum has been hidden. Most of the bullets would encounter nothing but hay and these would pass right through the bale and out the other side. But should a bullet chance to hit the platinum, it would ricochet at some angle. And if an enormous number of bullets were shot at the bale, so that the hidden nugget was hit often, ricochets in various directions could reveal the nugget's location and shape.

Rutherford and Geiger had sent many thousands of alpha-bullets through atoms; none had been turned aside (scattered) at an angle greater than a few degrees. These slight deflections could be explained on the assumption that an alpha-bullet, meeting an electron in its journey, had been affected by the electron's negative electricity.

In order to make this calculation and others like it, one had to know something about the mathematical theory of probability (the laws of chance). Rutherford was far from being a brilliant mathematician and in the past had tended to choose research problems which did not require much mathematical interpretation. Confronted with the problem of interpreting his alpha-scattering experiments, he had returned to school. As professor at the University of Manchester, he lectured to students—that was in the physics department. When some of the same students attended a mathematics class, they found Rutherford again, now a student, sitting beside them and industriously taking notes.

In the mathematics class he had learned how to calculate various degrees of probability. According to his calculations, there was a chance, a tiny chance, that an alpha-bullet, passing through a target of atoms, would meet one electron, another, another, and so on. The effect of these meetings would be to send the bullet off its course at a wide angle, perhaps as much as 45°. But the chance was slight indeed.

Yet this was the "small research" which Rutherford suggested for Ernest Marsden: he was to bombard atoms with alpha particles to see whether any were scattered at a wide angle and, as Rutherford said later, there was every reason to believe that Marsden would not find one. Still it should be tried.

And so Marsden went down to the laboratory's gloomy basement, where there were water pipes on the floor over which people tripped, and on the ceiling another pipe, just at head-banging level. Here a simple apparatus had been set up, according to Rutherford's specifications. It consisted of little more than a glass tube in which were enclosed a source of alpha particles, a target and a detector. The particles passed through a narrow slit to the target of thin metal foil (Marsden used gold), then continued on their way to hit a fluorescent screen, which acted as detector. An alpha particle hitting this screen produced a tiny, faint flash of light known as a "scintillation." Its position on the screen indicated how much, if at all, the bullet had been deflected by the atomic target. Now a variant of this apparatus was set up so that alpha particles scattered at angles of 45° and more also could be detected—should there be any.

Through a microscope Marsden would have to note thousands of the tiny, short-lived scintillations. It was difficult, tiring work. Before one began, it was necessary first to spend half an hour or so in the dark, waiting for one's eyes to adjust. Cups of "laboratory tea" would be drunk as the research workers sat in the dark, gossiping and trading jokes. Rutherford, on his daily rambles through the laboratory, often joined them. He lacked the patience himself for scintillation counting, having tried it once and, as he said, "damned vigorously and retired after two minutes." (Hans Geiger, on the other hand, was known as "a demon at the work." Later he invented the counter, named after him, which, by detecting the alpha particles electrically, released the human counter from these labors.) While Rutherford did not count scintillations, he was an active participant in the work at Manchester. Not only was the idea for the research his, as well as the scheme of attack, it was Rutherford who solved ticklish problems with apparatus and Rutherford who explained (in the lecture room as well as in the laboratory) what the research was aimed at. It was Rutherford also who raged when equipment broke down and who, when the research was going well, marched through the laboratory in triumph, roaring "Onward Christian Soldiers" (virtually the only piece of music he knew) out of tune.

Early in 1911, months after Marsden had completed the assigned experiment, a very triumphant Rutherford sought out Hans Geiger to tell him some news and to tell it, characteristically, with considerable drama: "Now," he said, "I know what the atom looks like!"

Contrary to all expectations, Marsden had found that out of the thousands of alpha particles he had tracked through a gold foil some, a very few, had been deflected at wide angles. Of these, one or two had been turned aside by more than 90°; they had come out of the target on the same side they had entered it. From calculations Rutherford was quite certain that these ricochets could not have been caused by a series of collisions between an oncoming alpha particle and electrons. What could have done it? J. J. Thomson's atomic model provided no clue.

At this point it would have been quite reasonable to conclude that Marsden had made a mistake. Or that some unknown factor, some contamination, had caused the bizarre result. Possibilities for error were numerous and hardly worth tracking down. Rutherford, who possessed a great talent for distinguishing the true leads turned up by experiment from the false, had not drawn this conclusion. He had taken Marsden's results seriously. That there was something in the atom which could bounce back the fast, massive alpha-bullets struck him as incredible, ". . . almost as incredible as if you fired a fifteen-inch shell at a piece of tissue paper and it came back and hit you." At once he began trying to find out what Marsden's results could mean. What had turned back the bullets?

Calculation told him that they must have encountered a tremendously strong electric field. Such a field could be produced by an electric charge concentrated into a very small space. An hypothesis took shape: the atom's positive electricity was not, as J. J. Thomson believed, a fluid evenly distributed throughout the atom. It was concentrated in a central massive core.

On the basis of this hypothesis, Rutherford set up a problem: Given a central electrical charge of so-much and a known quantity of alpha particles traveling toward it at a known speed, what would be the most probable scattering pattern to expect? About how many alphas would come close enough to the charged core to be scattered at an angle of 20°? 45°? 60°? 90°? After he had computed the answers, he compared them with Marsden's observations and with scattering experiments which had been done in the past. The agreement was close; he was on the right track. He looked up other experimental data, to find that it also checked with his hypothesis. Then he was sure, as he told Geiger, that he knew what the atom looked like. But the hypothesis must be tested in greater detail and, together with Marsden and Geiger, he designed new scattering experiments. Before the work was done, his two helpers had counted more than one million scintillations.

In May, 1911, Rutherford published his first paper on their joint findings and thus announced the discovery of the nucleus, as he later named the atom's charged core. On the basis of scattering experiments, he had been able to estimate its size: the nucleus was ten thousand times smaller than the atom—as small, comparatively, as a pinhead in an auditorium. Yet in the nucleus was found nearly all the atom's mass. Outside this minute, heavy speck at the center of the atom was space. There electrons were found in enough numbers to offset the positive electricity of the nucleus.

The atomic model which emerged from the work of Rutherford and others resembled a planetary system, for the force which binds planets to the sun obeys the same general form of law as the force which binds electrons to the nucleus. Both gravity and electricity decrease in strength with the square of distance. From this it follows that the particle-electron, attracted by the positive electricity of the nucleus, should move around it in the same way that a planet moves around the sun.

It was a compelling idea, an atomic world which duplicates in miniature the world of the sky. It was an idea which would form the basis for many important discoveries about matter. Still, it was but a beginning; there was, in fact, something fundamentally wrong with the solar-system model of the atom. What that was we will come to in the following chapter. First we will go backward in time to learn something more about the work and life of Ernest Rutherford, a man who dominated early atomic research, a man his associates described as "a force of nature," "a tribal leader," "a savage."

Rutherford was forty when he discovered the nucleus and already he had won the Nobel prize. This had been awarded him for the demonstration that radioactive atoms change spontaneously into atoms of a different nature, a discovery which helped to direct the attention of physicists to atomic structure. With the discovery of the nucleus, he revealed an essential fact about that structure.

Later his work took another dramatic turn as, following up his discovery of the nucleus, he was first to knock out a part of it and thus change one element to another by artificial means.

Speaking of this chain of triumphs, a physicist friend of Rutherford's told him once, "You are a lucky man. . . . Always on the crest of the wave!"

"Well," replied Rutherford, who was not known for modesty, "I made the wave, didn't I . . ."

This wave had begun to form soon after he left his native country, New Zealand, for England. Rutherford was twenty-four years old then, a large, dark, snub-nosed young man with definite opinions, which he uttered in a loud voice; a huge drive to get ahead in science; and no money at all.

In New Zealand, where he had received his education, he had won a scholarship to do postgraduate work at the Cavendish Laboratory, which was associated with Cambridge University in England. This was the first laboratory in the world to be de-

*Ernest Rutherford at the age of twenty-one, three years be-
fore his arrival at the Cavendish Laboratory.*

signed expressly for experimentation in physics, and its director
at that time was J. J. Thomson, whose hypothesis of atomic
structure Rutherford later would refute.

Soon after arriving in England, Rutherford paid his first call
on the man who would be in charge of his work, Professor

Thomson. Titles, including that of professor, meant very little to Rutherford, then and later. He detested pomposity and had little use for learned men who were interested solely in their own particular specialty. But Professor Thomson—or "J.J.," as everyone called him—was "not fossilized at all," Rutherford wrote in a letter home. He was friendly, "quite youthful still" (J.J. was forty at the time), and shaved badly.

Some others at the Cavendish did not welcome the New Zealander as warmly as did J. J. Thomson. Cambridge University had just been opened to graduate students from the British colonies like Rutherford, and many Cambridge natives resented "the intruders" with whom they had to share fellowships and desirable positions. At the Cavendish Laboratory, they had to share apparatus and instruments as well, and due to the great thrift of J.J. these were scarce. It was said that "the Cavendish man should, when preparing for an experiment, gather apparatus with his left hand and carry a drawn sword in his right."

Not only was the new man from New Zealand an outsider, he also had none of the marks by which an educated, class-conscious Englishman recognizes one of his peers. Rutherford's name meant nothing (his father owned and operated a sawmill); he was not particular about manners, dress, or speech. Finally, he came from Britain's youngest colony; by and large Englishmen regarded New Zealand as a decidedly backward place. It was known for the once warlike Maori tribe and for kiwi birds.

Soon after paying his call on J.J., Rutherford went to work. Rather than asking his superior to suggest a research project, as did many Cavendish students, he began one on his own initiative, first going on a major apparatus-collecting expedition, which undoubtedly did not increase his popularity. Then he built an instrument for detecting radio waves, a primitive wireless set which he had invented in New Zealand, not long after electromagnetic waves first were detected. Like Marconi, but even before him, Rutherford had seen a way to use the discovery.

J.J. showed interest in the project and some of the advanced research students, said Rutherford, "gnash[ed] their teeth with envy" and put little obstacles in the way of his work. Writing home, he said that there was one student "on whose chest I should like to dance a Maori war-dance."

He was burning for some success in physics. If his work were recognized, it would lead to an academic position. Then he would be able to repay the money he had borrowed for the voyage to England and most important marry the girl, Mary Newton, whom he had left behind in New Zealand.

To her he sent long letters, saying in one that he wished one of his scientific papers were a novel so that he might dedicate it to her, and in another that he sometimes went unshaved since in England, "I have not the inducement to keep smooth." He told her also that he was up against stiff competition at the Cavendish: "Among so many scientific bugs knocking about, one has a little difficulty in rising to the front." But in a notably short time this is just what he did. In Rutherford's life there were no years of struggle for recognition.

After only two weeks, J.J. recognized, as he later said, that his new student possessed "quite exceptional ability and driving power." He liked the forthright, unpretentious young man. And Rutherford proudly wrote to Mary Newton that he had been invited to speak before "J.J.'s pet society" and demonstrate his wireless. He was a success; other invitations followed. As he extended the range of his set to half a mile or so, he became quite well known around Cambridge. Research workers, once cool, now offered him their assistance and told him confidentially about their own projects. "So wags the world," was Rutherford's comment. To Mary Newton, he also reported breezily: "I am at present very useful when he [J.J.] is writing to various scientific pots as he can mention what his students are doing at the lab.

Some good startling effects with waves suit him down to the ground."

Rutherford himself was not so excited about his wireless invention. When J.J. asked him to join a study of X-rays he was planning, Rutherford put the wireless aside, even though he thought that it would be worth money someday. (He was correct and Marconi, who improved the one like it which he had invented, made the money.) Basic research of the kind J.J. was planning interested Rutherford. That was physics, he felt. Improving his wireless, that was merely engineering.

It was 1896 when Rutherford made this decision; Wilhelm Roentgen had just announced his discovery of rays which, because they were an unknown quantity, he had named "X." Unlike many important discoveries, this one had caused an immediate sensation. With his X-rays, Roentgen had made photographs of the bones in his hand and of some metal pieces inside a wooden box. Scientists everywhere were duplicating his work, "to see their own bones," as Rutherford put it. The Cavendish was buzzing with excitement over the discovery. "The great object is to find the theory of the matter before anyone else," Rutherford had told Mary Newton, "for nearly every professor in Europe is now on the warpath."

As it turned out, Rutherford and J.J. did *not* find the theory of the matter; their work was important, however, because they learned about a property of X-rays which was decisive for later atomic research. This property had been detected by Roentgen when he noted that X-rays impart electrical conductivity to a gas. Now Rutherford and J.J. found out why: X-rays sent through a gas, such as air, cause the formation of ions (particles carrying an electric charge). The ionizing property would assist J. J. Thomson in his measurements of the electron's charge and mass; study of the same property would lead Rutherford to an understanding of radioactivity.

But these achievements came later, when the two men no longer were working together. Their collaboration lasted for six months and during that time Rutherford did most of the manual labor. J.J. planned but did not perform experiments. His son George Thomson, (also a physicist and a Nobel prize winner) has explained why: J.J. "was surprisingly clumsy . . . and though he could diagnose the faults of an apparatus with uncanny accuracy it was just as well not to let him handle it."

Rutherford, on the other hand, was exceptionally skillful. His setups were crude, it is true; they have been described as looking "like nothing on earth." But they were suited perfectly to the job at hand. "The minimum of fuss went with the minimum chance of error," said a colleague. Rutherford's method of attack was simple; the desired result was obtained with little effort and confusion. Another colleague of his, A. S. Russell, put it very well: "With one movement from afar, Rutherford so to speak threaded the needle the first time." The flair for experimental design, together with an ability to pick out the one significant fact from a mass of confusing detail—these were Rutherford's greatest talents as physicist. Both were characterized by simplicity.

After the study of X-rays' ionizing property was complete, Rutherford did not return to his wireless set. Instead, he used the techniques he had mastered to start independent research on the ionizing property of another form of radiation—ultraviolet light. That done, he chose to investigate in the same way yet another radiation, the one emitted by the radioactive element uranium. In his career, this was a turning point.

Radioactivity had been discovered just a few months after X-rays; indeed Henri Becquerel was seeking more information about X-rays when, partly by accident, he found evidence of the penetrating rays emitted by uranium. When he announced his discovery, physicists were so busy with the study of X-rays that few—Marie Curie was a notable exception—chose to follow up his work. This situation changed dramatically two years later, in 1898, when she reported her discovery of new radioactive substances, one of them, radium, being several million times more active than uranium. Then physicists and also chemists became intensely interested and soon there was a group of scientists who sometimes called themselves "the radioactive people." In physics radioactivity had become the fashion, just as later on the atom would be, and then the nucleus.

And in every case, Rutherford was in at the start: he had chosen the right problem to study at the right time. Thus with J.J. he had developed just the techniques necessary for learning more about radioactivity and was prepared to exploit the new discovery. Very soon, as he studied the ionization property of the rays emitted by uranium, he learned something important: two distinct types of radiation were given off by the element. One, which he named "alpha" radiation, was easily absorbed by matter; the other, "beta" radiation, was far more penetrating. (A third and even more penetrating ray, "gamma," was identified a year later by another physicist.) Because these rays were unknowns, like those Roentgen had called "X," Rutherford had chosen to call them by the first letters of the Greek alphabet. He hoped to learn their true identity.

While in the midst of this research on radioactivity, Rutherford was offered a good academic position, but one which presented a problem. The offer came from McGill University, in Montreal, Canada, where there was an opening for a professor of physics. On the face of it, this was exactly the opportunity Rutherford had been waiting for: he would be a professor (at the age of twenty-seven) and have a salary large enough so that he would be able to marry. And at McGill, Rutherford wrote home, there was "a swell lab." Still, he hesitated. It meant leaving the Cavendish; it meant leaving the part of the world where things were happening in physics.

At this time there were fewer than four hundred physicists in the world; and nearly every member of this small group was to be found at a university in the British Isles or on the European continent: the universities of Cambridge, Berlin, Göttingen, Paris, and so forth, where they taught and pursued research. These universities were not very distant from one another and physicists could see one another quite often to trade information turned up by experiment, compare their different interpretations of it, and ask each other questions. Such meetings were important to all. Then, as today, a person in scientific research was dependent upon a rapid flow of information if he were to profit from the work of others—and avoid duplicating it. One could not keep up with research simply by reading reports of it published in scientific journals. These reports did not appear until weeks, even months, after the work had been completed. To keep in close touch, one had to be on the scene.

But at McGill, Rutherford would be a long ocean voyage away from this activity. In Canada and in the United States there was, compared to Europe, almost no physics research going on—just a few men, separated from each other by great distances and working in virtual isolation. The scientific journals arrived late; almost all of them originated in Europe and Great Britain. To leave England for Canada, Rutherford once told a friend, was to leave "the Physical World"; and although he meant the world of physics and not life itself, the two, for someone like him, were not very different.

It was largely due to J. J. Thomson's influence that the professorship at McGill had been offered to Rutherford and the young man suspected that if he stayed on at the Cavendish and waited, J.J. would help him to get another position, one which was not so far from the physical world. Therefore it might be wise to refuse the McGill offer. Rutherford weighed the pros and cons, then made his decision—for McGill. It was not like him to wait patiently and live on hopes; as well as being in love, he was practical. Once he *was* a professor, at McGill, other universities would be more likely to consider him when they had an opening.

And so he wrote to Mary Newton, "Rejoice with me, my dear girl . . . for matrimony is looming in the distance." (They were married two years later.) In the same letter he told her, "I am only a kid for such a position. . . . It sounds rather comic to myself to have to supervise the work of other men, but I hope I will get along all right. There are about four men doing research in the lab., some of them as old as myself, so I will have to carry it off somehow."

A few years before the turn of the century, Rutherford pulled out of the Cavendish and sailed (via New Zealand) for Canada. He returned to the colonies, from which he had come. At McGill, as well as learning the duties of professor and how to supervise a laboratory, he would continue his research on radioactivity, attacking problems which others among the radioactive people, including Becquerel and the Curies, also were trying to solve. Despite the handicap of distance he intended, he said, "to keep in the race."

CHAPTER TWO

Ernest Rutherford:

Radioactivity

Well, it's a great life.

—Lord Rutherford

IN HIS LETTER to Mary Newton about the new position at McGill, Rutherford had sounded uncertain about his ability to supervise other men. But once on the scene he rapidly organized McGill's physics laboratory for research on radioactivity and before very long he was heard praising the work of one of his associates, a man eight years older than himself, with the words, "Good *boy!*" Rutherford's colorful and long career as a leader of research had begun.

He was a competitive man: science was "a race" with "other sprinters . . . always on my track." But the object of the race was not the prize (although he enjoyed them). Rutherford wanted passionately, even desperately, to find out, to learn something new. Thus in the role of professor he was apt to talk about unsolved problems—current work in physics, including his own —rather than work which had been done in the past. As a result his students remained ignorant of much that was worth knowing, as some have testified. What is more, they were pressed into the laboratory to work on Rutherford's problems, sometimes when they had received little or no previous training for such work. The research problems came thick and fast—too fast, some people said. "Get on with it," Rutherford was always telling them.

Those who were his students at McGill and at other universities have not remained silent about the nature of their appren-
ticeship to Rutherford, nor are they uncritical: "I never," says
one, "met anyone who could get quite so angry over trivial things
as Rutherford—though he always apologized for it." But above
all these students speak in fondness for the man "who had none
of the meaner faults"; who was "just as willing to attend to the
youngest student and if possible learn from him as . . . to listen
to any recognized scientific authority"; who made them feel as if
"we were living very near the center of the scientific universe."
For Rutherford science was a race; but he was not a lone runner.

One of his students, H. R. Robinson, has described a long
Saturday afternoon he spent with the New Zealander in an attempt to purify, with a few dregs of liquid air, a sample of a
radioactive element with which they hoped to work. They had
not been successful, due to a false move on Rutherford's part,
which was accompanied by the remark, "Well, it's a good job *I*
did that and not you." As professor and student together cleaned
up the mess they had made in the laboratory, Robinson was feeling a bit disgruntled over his wasted Saturday afternoon. Not so
Rutherford, who told him, sucking contentedly on his pipe,
". . . You know, I *am* sorry for the poor fellows that haven't
got labs to work in."

This incident occurred after Rutherford had left McGill, after
his research there had advanced him to the summit of his profession. He was soon to become "Sir Ernest" and then "Lord
Rutherford." Despite his isolation from the physical world in
Canada, he had indeed kept in the race.

His research at McGill had begun where he left it in England,
with the alpha and beta rays. Marie and Pierre Curie, Henri
Becquerel, and other "sprinters" already had uncovered the identity of "his" beta ray. It was not a form of radiation, as had been

thought, but consisted of electrons moving at a great speed, almost as fast as the speed of light. Then what was the alpha ray? Rutherford established that it too was composed of particles moving at high speed. They were far more massive than the beta-electrons and carried a positive charge. But again, what were they? He suspected their identity when he learned that traces of the element helium always were found in minerals which contain radioactive elements. But it was ten years before he was able to prove decisively—in the spectrum experiment done at Manchester—that the alpha particles were indeed positively charged helium atoms (helium nuclei we would say today).

Study of radioactivity had raised a host of other problems in addition to "What is the alpha?" The most fundamental of these was the question, "Where do the great energies of radioactivity come from?" That the atom itself contained these great energies was not understood; it was generally assumed that radioactive substances had absorbed energy in some way from their external environment. This assumption blocked the way to a clear understanding of radioactivity. Scientists were accumulating facts about the process, such as the properties of alpha and beta rays. They were identifying other radioactive substances in addition to those which Marie Curie had detected. But what was the connection between the emitted rays and the newly discovered substances? And were these substances compounds, combinations of known atomic varieties, or were they new elements, new varieties of atom? The experimental discoveries did not fit into a meaningful pattern. There was, in other words, no general theory of radioactivity. This would be the contribution of Rutherford and of the young collaborator he found at McGill, Frederick Soddy.

Soddy was twenty-three years old when, in 1900, he began to work with Rutherford. He also had come from England to teach at McGill, to teach chemistry. Rutherford needed a chemist for his radioactive research, which involved such procedures as the chemical isolation of an element from other substances. Just a few weeks after coming to McGill, said Soddy, "Rutherford . . . got me. I abandoned all to follow him and for more than two years scientific life became hectic to a degree rare in the lifetime of an individual."

Soddy discovered that the New Zealander had rather a low opinion of the science of chemistry, as indeed of all other natural sciences, barring only physics. That, according to Rutherford, was in a class by itself. Physics was concerned with large matters, with universals; other sciences, he felt, merely studied the details, the local variations. Although he needed Soddy's knowledge of "details" for the research (and profited greatly from Soddy's theoretical abilities), he delighted in proving to his partner that a physicist could beat a chemist at his own game.

Once he took time out to demonstrate, for Soddy's benefit, how it was possible to isolate an element without using any chemical techniques at all. First he added thorium dioxide, a white earthy substance, to water, gallons of water. Then, grimly determined, he shook them up together, an exhausting process. Finally, he boiled down *all* the water. Hugely pleased with himself, he showed Soddy the end result of his physical labor—a minute quantity of the new substance which Soddy and he recently had discovered—thorium X.

During the hectic collaboration between these two, they published a series of papers which, together, constitute their theory of radioactivity. The first experimental clue to the theory had been turned up when one of Rutherford's students, given the problem of measuring the ionizing property of the radioactive element thorium (which gave its degree of radioactivity), had run into difficulties. His electroscope gave a different intensity of ionization at different times; he could not arrive at a definite measurement. And curiously enough, the different intensities seemed to depend on whether the door to the laboratory was closed or open.

At this point Rutherford became very interested in the problem. Before long he had explained the capricious measurements with the discovery that the element thorium emits a radioactive gas (now called "thoron"). When the laboratory door was closed, the gas hovered over the element, adding its radioactivity to that of thorium; but when the door was open, the gas was blown about the laboratory by air drafts. Said the physicist P. M. S. Blackett, commenting on this discovery: "May every young scientist remember . . . and not fail to keep his eyes open for the possibility that an irritating failure of his apparatus to give consistent results may once or twice in a lifetime conceal an important discovery."

Further investigation of this discovery revealed that thoron was not formed *directly* from thorium: there was an intermediate substance involved, the thorium X already mentioned. Thus thorium transformed itself to thorium X and thorium X to thoron. Then perhaps all radioactive atoms, by expelling an alpha or beta particle, spontaneously transformed themselves into atoms of a different character—new elements. And these in turn broke down, to form still other new elements. The energy of radioactivity was that of the atom itself as it changed, or to use the technical word, decayed.

This was the pattern into which Rutherford and Soddy fit the many facts known about radioactivity. There were, they established, three main families of radioactive elements, one beginning with thorium, one with actinium, one with uranium. All other radioactive elements were offspring, decay products, of these three elements. Radium, for example, was one of the decay products of uranium.

There was one very significant omission in Soddy and Rutherford's theory of radioactivity. It said nothing of the time when any particular radioactive atom would expel a particle and thus transform itself. Whatever triggered this process was unknown. Attempts had been made to accelerate the decay or slow it down; they had failed. Heat, cold, other external conditions did not affect the rate of decay. The rate also did not change when a radioactive element joined with another element to form a chemical compound. And the age of the atom had nothing to do with it either. The decay rate of radium was the same whether the radium atom had existed for a thousand years or had been newly formed from the decay of a heavier atom.

It was clear that radioactive decay was due to an internal change of the atom (a breaking up of the nucleus, it was later established), but what made this change occur was unknown. That was why the theory of radioactivity did not make a prediction as to the future behavior of any single atom. In working out a rate of decay, Rutherford and Soddy had had to employ statistical techniques similar, in a sense, to those used by an insurance company to estimate the length of life. There being no way to tell how long a particular individual will live, the insurance company bases its predictions on study of the lifespan of millions of individuals. In the same fashion, Rutherford and Soddy worked out decay rates for the different species of radioactive atoms. For example, they established that radium has a half-life of 1600 years. This means that after 1600 years, half of a large quantity of radium atoms will have decayed to form radon, the next decay product in the uranium family. Theory gave the various decay rates precisely, but only for a vast quantity of similar atoms. Of the individual, it said nothing.

At the time when these decay rates were established, it was assumed that experiment someday would answer the question, "What triggers the atomic change of radioactivity?" Today physicists have learned not to seek an answer to this question. They point to the theory of Rutherford and Soddy as an early indication of the atomic physics which, at that time, lay far in the future, a physics which does not predict the future of individual atoms but only of a large group composed of identical members. Later on in this book, we will be concerned with that aspect of Rutherford and Soddy's work.

The theory of radioactivity which the two men proposed over half a century ago today stands largely unchanged. Much has been added to it; but nothing has had to be eliminated. Today the ideas it introduced—intra-atomic energy, transformation of the atom—are commonplaces. In 1902, however, they sounded to some ears rather farfetched, unlikely. As in the years following 1902 Rutherford and others accumulated more and more evidence which backed the new ideas, they were accepted gradually by the community of scientists. But at first there were doubters, particularly among chemists and the older generation of physicists—even the Curies.

"Everyone wanted to jump on me in those days," said Rutherford. One staunch opponent was the great English physicist Lord Kelvin, who at this time was over eighty years old. Long before, he had calculated the age of the earth, basing his arithmetic on the laws for the cooling of a body. Now Rutherford was saying that if one took the intra-atomic energies released in radioactivity into account, it would have taken the earth far longer to cool than Kelvin had estimated. Rutherford had arrived at a new figure for the earth's age, using a new technique. He had measured the amount of helium in a sample of pitchblende, a mineral source of radium and uranium. Since he knew the rate at which the uranium family decays, giving off alpha particles (which he assumed to be helium), he could calculate how long the mineral had existed in a compacted form.

Defending his own work, Kelvin claimed that the new theory of radioactivity was entirely wrong. Radium was not an element, he said; it was a molecular compound of lead and helium, and had acquired its energy by absorbing "ethereal waves."

Rutherford met his ancient adversary face to face when he

paid a visit to England to attend a conference of physicists. "Lord Kelvin," he wrote back to his wife, "has talked radium most of the day, and I admire his confidence in talking about a subject of which he has taken the trouble to learn so little."

At that time scientists often played parlor games with the newly discovered radioactive elements and at a party held during the conference Rutherford had demonstrated how radium could make a phosphorescent material glow in the dark. Lord Kelvin watched. "He was tremendously delighted," Rutherford reported home, adding somewhat paternally that the old man had "gone to bed happy with a few small phosphorescent things I gave him."

While a venerable physicist's objections to his work did not seem to ruffle the young New Zealander's great self-confidence, certain other criticisms infuriated him, as his letters reveal. Quite truthfully, Rutherford could have declared that "some of his best friends were chemists." One of these was Bertram B. Boltwood, of Yale University, who also was studying radioactivity. The two men wrote often to one another, and once Rutherford told his chemist friend that he had been reading articles in the scientific journals which criticized his theory for not being sufficiently grounded in experiment. The writers of these articles were "damned fools, whom I think must once have been chemists," Rutherford fumed. "Excuse me—no personal reference," he added, as he remembered that Boltwood *was* a chemist, and then went on to say that the article writers did not have "the faintest notion that the . . . theory has as much evidence in support of it as the kinetic theory . . . and a jolly sight more than the electromagnetic theory, which they all swallow as the eternal verities."

Still, formal recognition of the new theory was not very long in coming; Rutherford did not have to wait twenty years or longer for the prizes as have a number of other physicists. Never one to look down his nose at an award, particularly if a gold medal went with it, he was happy to learn in 1908 that his "investigations in regard to the decay of elements and . . . the chemistry of radio-active substances" had won him the Nobel prize. He wrote to his mother that it was "very acceptable, both as regards honour and cash."

Less acceptable to a man of his views was the fact that the Nobel prize had been given to him for work not in physics, but in chemistry.° In the customary speech given by each Nobel prize winner he remarked dryly that he had observed many transfor-mations in his radioactive work but never had seen one quite so rapid as his own, from physicist to chemist.

Rutherford was thirty-five years old when he returned to England, where he would spend the remainder of his life. He had stayed at McGill until a professorship was offered him which met certain requirements, the main ones being proximity to the European centers of physics and good laboratory facilities. The University of Manchester offered both. It was there, as we have seen, that the nucleus was discovered.

The year was 1911. In 1914, World War I would begin and almost all research at Manchester would come to a stop. Mars-den would fight on the side of England; Geiger for Germany. One of Rutherford's most promising students, H. G. J. Moseley, would be killed in the Gallipoli campaign.

The short period before 1914 and after 1911 marked a golden age at Manchester University. As experiments were done tracing out the consequences of the new atomic model, with its nucleus core, discoveries were made, it is said, "at the rate of about one a

---

° At this time research involving the elements was considered chemistry, which is why Rutherford was awarded the Nobel prize in that science. Today the situation is quite different. It is sometimes said, only half jok-ingly, that chemistry has become a branch of physics.

week." The tortured melody of "Onward Christian Soldiers" rang through the laboratory as the big, red-faced New Zealander with the shaggy mustache made his daily rounds.

During the afternoon break for tea Rutherford and his research team would discuss the current work and decide what should be done next. Ideas were exchanged very freely at these daily get-togethers. There was little of the usual worry that if one gave voice to an idea another man, seeing a way to develop it further—and faster—would get his results into print sooner, and thus receive the credit. Such fears were unnecessary. The discovery of the nucleus had opened up a new world for exploration. Everyone was having good ideas. It was, said Marsden, "almost immaterial who did the work and published the results. There were good pickings for all and no one jostled for a particularly spicy tidbit."

Outside the Manchester laboratory, however, the situation was very different. Few physicists were exploring implications of the new atomic model. To a certain extent Rutherford himself may have been responsible for this indifference to his work, for in his publications he had failed to show just how important it was. The nuclear atom had been presented as the conclusion to which his scattering experiments had led, but only as that. He had not made a strong case for it; he had not, for example, attempted to explain known chemical properties of the elements on the basis of the new model. And J. J. Thomson's atom, *did* account for some chemical properties of the elements. Those physicists who were following the subject of alpha-particle scattering knew about the evidence Rutherford had found for a nuclear atom. But not many physicists were following this subject.

At that time very few were interested in research of this sort. Before trustworthy conclusions about the atom could be reached, one would have to have a vast quantity of experimental evidence. This did not exist (or so it was thought) and

ULLSTEIN

Ernest Rutherford, flanked by two "boys," E. T. S. Walton on the left; J. D. Cockcroft on the right. The picture was taken in the early 1930's when Cockcroft and Walton, working under Rutherford at the Cavendish, achieved the first transmutation of the nucleus by means of artificially accelerated atomic particles.

the prospect of obtaining it was not good. According to E. N. da C. Andrade, physicists of those days considered the atom almost as hard to reach experimentally as a planet. In 1911, it was futile to ask, "Is there life on other planets?" It appeared about as

profitless to ask, "What is the atom?" There just was not enough
to go on.

Then, in 1913, something happened which dramatically
changed this state of affairs: one of Rutherford's students solved
a fundamental problem and, in doing so, demonstrated that physicists were wrong about the atom's inaccessibility. There *was* a
way to learn of its structure and behavior, learn of it accurately
and in detail. This student's work opened the way toward a quantitative atomic science—the precision tool we have today. In the
years following 1913, atomic research would become the principal concern of physics; once again, Rutherford would be on top
of the wave—thanks to the work of his student.

The student's name was Niels Bohr. He had seen Rutherford
for the first time at a dinner party held at Cambridge University,
the annual Cavendish dinner, a traditionally boisterous occasion
when J. J. Thomson's present and past research students met together to give speeches, eat, drink, joke and sing. At that time
Bohr was a student of J.J.'s. Like Rutherford, sixteen years before, he had been attracted from his native country (Denmark in
Bohr's case) by the great work which issued from the Cavendish
Laboratory.

At the ceremonial dinner, the young Dane heard Rutherford's
big voice extolling "the most wonderful instrument in scientific
history." He was talking about C. T. R. Wilson's new cloud
chamber, which made the ionization produced by X-rays and
atomic particles visible to the eye. "The charm and power" of the
New Zealander's personality made a deep impression on him,
Bohr said later.

Before very long, Rutherford, having returned to Manchester,
mentioned in a letter to his friend Boltwood that "Bohr, a Dane,
has pulled out of Cambridge and turned up here to get some experience in radioactive work."

The young man had come to Manchester in the early spring of

1912, just a few months after the discovery of the nucleus. His fellow research students were excitedly pursuing its consequences. But in spite of Rutherford's pressure "to get on with it" in the laboratory, in spite of the fact that Bohr liked to experiment and had come to Manchester expecting to do exactly that, he did not help himself to some of the "good pickings." Rather than working out possibilities which the nuclear atom implied, he was thinking about an impossibility which it also implied. As we have suggested earlier, there was something wrong with the atomic model based on the discovery of the nucleus, the model which depicted the atom as a miniature solar system.

According to this model, the electron was attracted to the nucleus of opposite electrical charge. Therefore the electron would move; it would move, like the planets, in an elliptical orbit around the nucleus-sun. But a moving electron was impossible. Why? Because, according to the laws of electricity, a moving charge must produce electromagnetic radiation, light. The electron, always in motion, would produce radiation; all atoms would emit light at all times. But matter under ordinary circumstances does not glow with light.

This was one flaw of the planetary atomic model; there was another, closely associated with it. A moving electron, to repeat, must produce radiation. In doing so, it would lose energy and therefore spiral into the nucleus, much as a satellite, feeling the drag of air, spirals back to earth. Weeks or months pass before a satellite plunges into the earth; the electron would fall into the nucleus in a fraction of a second. Then there should be no such thing as an atom; there should be only a nucleus. The model which represented the atom's structure simultaneously denied the possibility of that structure. It was this problem which Niels Bohr would solve, and in solving it he founded the science which evolved into today's atomic physics.

In a later chapter we will return to Bohr, to the solution he found and to the role in his life which Rutherford played. Now let us put aside the problem which the nuclear atom posed, because the key to its solution lies in an idea which has not yet been introduced, an idea called "the quantum theory." This theory was thirteen years old when Bohr applied it to the case of the atom, thus showing its relevance to the fundamental structure of matter.

In the next chapter we will go back in time to speak of Max Planck, who in a persistent attempt to solve a problem concerning radiation advanced the idea of the quantum. Perhaps it would be more accurate to say that he backed into the idea, for as we will see his discovery was made in a curious fashion. After leaving Planck, we will not return to Manchester and Niels Bohr, but first will introduce a man who enlarged Planck's quantum theory, a man who will make more than one appearance in these pages—Albert Einstein.

This retreat in time, this remove from England to Germany, where Planck and Einstein were born, will take us far away not only from the atomic problem which we have raised but from the kind of physics at which Ernest Rutherford excelled. We will be concerned with physicists who do not as a rule perform experiments—theoretical physicists—and with such questions as "What is a physical theory?" "How do men who construct such theories go about their work?" When at length we return to Niels Bohr, we will be in a position to see how he drew upon the work of Planck and Einstein and how his atomic theory then was broadened and changed, to become the one which is used today.

CHAPTER THREE

Max Planck: Pursuit of an "Absolute,"
the Entropy Law

> *The originator of a new concept . . . finds, as a rule, that it is much more difficult to find out why other people do not understand him than it was to discover the new truths.*
>
> —Hermann von Helmholtz

At about the time that Ernest Rutherford was developing techniques for probing the constitution of matter, other experimenters were improving techniques for the measurement of radiation. They developed an almost perfect light emitter, that is, a material body which when heated to high temperatures sends out, within certain limits, radiation of all possible wavelengths. This broadest possible spectrum of light, known as "black-body radiation," was used as a standard in the development of various sorts of lamps.

Scientists who were concerned with knowledge for its own sake, apart from its practical applications, also were interested in the black-body spectrum. The different colors of a spectrum and their varying brightness give the energies which are present in the radiation. In the case of the black body, the radiation spectrum is not influenced by factors due to the substance of the emitting body or the condition of its surface. The black-body spectrum, then, is a pure, an ideal, case. If one could describe with physics the energy distribution of the ideal case, then one would learn something about the radiation process in all cases. For this reason a number of physicists chose to work on the black-body problem. Starting with assumptions based on theories of heat and light then current, they tried to deduce a formula which described the distribution of energies in the black-body spectrum. These efforts failed and their failure became known as "the ultraviolet catastrophe." For logical application of theory to this particular case led to a formula which did not agree with experiment in the ultraviolet region of the spectrum.

The conclusion of theory was that a black body, heated to high temperatures, would send out an infinite amount of energy in the high frequencies,° which is to say, wavelengths in the ultraviolet and beyond. If nature actually did behave as this formula predicted, a lump of coal burning in the furnace, or any other radiant piece of matter, would emit energy in a burst of dangerous short-wave radiation.

In the black-body case, just as in the case of the atom, physics predicted catastrophe—a catastrophe which did not in fact occur. And both predictions were due to the same assumption, the idea that "nature makes no jumps," that energy is continuous (unbroken). Now virtually every natural change which man observes directly—that is, with his senses—*is* continuous. The pendulum, swinging freely in the air, comes to rest gradually and smoothly; its motion does not subside in stops and starts. It is not hard to understand that the idea of continuity in nature at one time appeared self-evident, true beyond question. Today we know otherwise; we know that the subatomic region of nature differs in essential ways from the objects and happenings which we observe directly. The physicist has learned to work

° The frequency of a wave is the number of times its crests (or troughs) pass a point during a given period of time. The crests of a short wave pass the point more often—that is, with higher frequency—than do those of a long wave.

with ideas which do not correspond with his ordinary experience, ideas which contradict that experience.

This radical change in science began with the formula which correctly described the energies of black-body radiation, denying that those energies were continuous. Max Planck, the man who succeeded in writing the formula, and by doing so introduced the quantum theory, did not make this denial knowingly. He did not say to himself, "Attempts to write a formula for black-body radiation have failed. There must be something wrong with our fundamental assumptions about nature. Suppose I try a different assumption, one that contradicts the evidence of my senses and the science of my time which sums up that evidence. . . ." No, Planck's discovery, like many others in science, was due at least partly to a fortunate accident.

Planck was like a man who, before the discovery of fire, wanted to find the best ways to bore holes, who for months, years, even decades, bored holes in every material he could find in every conceivable fashion and in doing so chanced to discover fire. In other words, Planck was pursuing a line of thought, pursuing it systematically, diligently, patiently. To others it looked as if the route he was following could not lead to anything important. But there is no way to foresee a great discovery before it is made; and it chanced that Planck's scrupulous and dogged pursuit brought him to the place where "fire" *could* be discovered. In the next chapters we will see how this happened.

Max Planck began to love physics when he was a boy. In school, in the *"Gymnasium,"* as it is called in Germany, he learned about a law of physics and never forgot that day. "Imagine," his teacher said, "a workman who lifts a heavy stone block, and struggling under its weight, raises it to the roof of a building. The energy of this work does not disappear. Perhaps one day, years later, the stone becomes loose. Down it falls on the head of someone below."

The boy Planck was struck by this example of the law of energy conservation almost as hard as the "someone" who had been struck by the illustrative stone block. It dawned on him then that the world was not complicated beyond man's ability to understand it. Amid seemingly endless complexity and variation, the human mind could distinguish order, could recognize law. That man's ability to reason was equal to this giant work seemed to Planck a miracle; the work itself he called "sublime," conceiving it as the discovery of pure and absolute truth. He would, he decided, become a physicist.

When he was seventeen years old and ready to enter the university, Planck sought out the head of the physics department and told the professor of his ambition. The response was not encouraging: "Physics is a branch of knowledge that is just about complete," the professor said drearily. "The important discoveries, all of them, have been made. It is hardly worth entering physics anymore."

This was the year 1875. The physics which had begun some two hundred years earlier with Isaac Newton's laws of motion had been extended so that heat, sound, electricity, light were understood in terms of models which operated in harmony with Newton's laws. Physicists had come to think of the universe as a giant mechanism whose basic workings were known to them. The discoveries which would dispel this illusion had not yet been made. Ultimately, due to these, the physics of 1875—which today is called "Newtonian" or "classical"—would be revealed as only a partial understanding. But when Planck spoke to the professor, this physics was the only one and in the opinion of most of its practitioners it was the last word.

For someone who hoped to discover something important the prospects were gloomy. Thus Planck, at the outset of his career, was greeted with disappointment. There would be others. In Planck's life there was no early recognition, no rolling wave of triumphs.

Max Planck in 1915. Three years later he was awarded the Nobel prize for the work on black-body radiation.

And there was no laboratory, no collaborator or research team of young men. Unlike Rutherford, Planck worked alone and except as raw material for theory he was not interested in experiment. It is rumored in fact that never in his life did he perform an experiment; and although this story probably is false, it is not very false.

As a physicist Planck was distinctly unlike Rutherford. Planck also was most definitely not a carefree, unrepressed, noisy sort of person. When he spoke, in a quiet voice, his words were carefully chosen. His manner was reserved and formal, his clothing was dark, his shirts very starched.

Still, like Rutherford, Planck was an energetic, hard-working, dedicated scientist. The discouraging remarks of the first authority on physics he consulted did not deter him from entering that science. What is more, he would choose to concentrate on a part of physics, thermodynamics, in which others had lost interest, thinking it was complete, finished. And within this part of physics he would devote himself to an idea in which, a colleague said, "literally nobody at all had any interest whatever."

It was not Planck's design to lose himself in obscurity; quite the contrary, he hoped to make a mark in science. He had decided against a career in music, much though he loved it, because he did not think that his ability was first-rate. He did not want to be merely a "good" composer. In choosing physics, then, he must have thought that he could do work of the highest quality. Like the men of his family, who were known for their self-discipline and lofty standards, he set his goal high just as, when he went mountain climbing, he singled out one of the highest peaks and climbed it, probably not so much for sport or exercise as for a sense of solitary, disciplined achievement. In his eighties, Planck still was climbing some of the higher Alps.

The snow-capped Bavarian Alps are just outside Munich, the south German city where Planck grew up and where Albert Einstein (who was twenty-one years younger than Planck) also spent his youth. As boys, these two knew the same black-and-white mountain forests of birch and evergreen; the same clear mountain lakes; the same music, which filled Munich, its taverns and beer gardens as well as its opera houses and concert halls. But while Einstein's father, the owner of a very small and unsuccessful business, ranked as a "nobody" in German society, Planck's father, a university professor, definitely was considered "somebody."

In Germany at this time only princes and barons were accorded more respect than professors; and the professor's family shared the glory. Should his wife—she had a title, too: *Frau Professor*—enter a shop, the clerk at once would stop what he was doing, and turning from waiting customers, overwhelm her with attention. Or suppose she attended an informal *Kaffeeklatsch* where the ladies of her circle met to gossip and consume creamy pastries. Among those present, there might be a woman much older than the *Frau Professor*, but if the husband of this woman was less than professor, instantly she would rise to her feet, old though she might be, and offer the *Frau Professor* her chair.

The world in which Max Planck grew up was one of distinct social gradations; it also was one of lofty ideals. The Planck family, in the Prussian tradition, felt deep reverence for the state. To rule, they believed, was the kaiser's absolute right; their obligation, also absolute, was to work, sacrifice, obey. From his family Planck also inherited a profound respect for law and justice. Several generations of Plancks had been scholars and administrators of law; they were known as just, incorruptible men.

As well as being born into different parts of German society, Planck and Einstein responded differently to its institutions—school, for example. Einstein, in his Munich *Gymnasium*, rebelled against regimentation, against dictatorial schoolmasters and the endless study of Latin and Greek; Planck, in his *Gymnasium*, was inspired. He accepted the teacher's absolute rule; the lack of freedom, even to choose a desk; the shame if one did not know the answer to a question. He enjoyed Greek and Latin so much that he considered specializing in language studies. And while Einstein in school already was questioning the whole structure of physics, Planck received its laws as absolute truths.

When it was time to enter a university he chose the one in his home city, Munich, where his father taught law and where many of the other professors were his father's friends. There, after his uninspiring reception by the head of the physics department, he experienced another disappointment. The ideas of physics attracted him, not the experiments; he wished to study theoretical physics, but the University of Munich, like most universities at that time, had no such department. Physicists were not divided as they are today into two main groups: on the one hand those who, primarily, design and perform experiments; on the other, those who generalize the results of many experiments into laws and fit the laws into a structure of theory. At the University of Munich there was a department of mathematics where the student could learn to use the symbols which may be applied to physics as to anything else, *if one knows how,* and there was a physics department where the students repeated experiments which had led to major discoveries. There was no one who taught the construction of theory. Although Planck said later, quite characteristically, that he retained his Munich professors "in reverent memory," he did not find what he wanted at Munich and after a few years he decided to transfer to another university.

Students frequently wandered from one university to another in those days, for once out of the *Gymnasium* there were few restrictions on their freedom. In the university one was not assigned to a class or required to attend it and there were no examinations except a final one to qualify for the doctor's degree (the B.A. and M.A. were not given by most European universities). Also there were no dormitories. A student lived in a room of his own and when he felt like going to another university— where there was a well-known professor perhaps, a good laboratory, or winter sports in the neighborhood—he moved.

Planck chose to transfer to the university in Germany's capital city, Berlin, where army officers and government officials as well as students attended lectures of the celebrated professors. (Von

Treitschke, professor of history at this time, was telling his large audience: "Our age is the age of iron; and if the strong vanquish the weak, it is the law of life.")

The University of Berlin represented the very summit of German science. Not only were German wanderers like Planck attracted to Berlin but also graduate students from other European countries and from the United States. The king of this mountain was Hermann von Helmholtz, professor of physics, whose work had played a decisive part in establishing the law which had so impressed the schoolboy Planck—the law of energy conservation. In Germany, where professors were very important, von Helmholtz ranked as the greatest of the professors. It has been said that "next to Bismarck and the old Emperor, he was at that time the most illustrious man in the German Empire." He had been given the awesome title *Exzellenz* and his colleagues always used it, greeting him with deep bows. So dignified was von Helmholtz, so awe-inspiring he looked with his huge head, his towering brow marked by furrows and a distended vein, that one of his students compared him to Wotan, father of the gods of Valhalla.

But to Max Planck's surprise and disappointment the lectures of this great physicist were boring. "Wotan" spoke very slowly and so softly that he barely was audible. He halted again and again to consult his notes. The numbers he wrote on the blackboard were so tiny that the audience often could not make them out, and frequently he made mistakes. Apparently, the great mind of von Helmholtz was busy with extracurricular matters.

One day a tall army officer with a big cigar in his mouth was seen entering the building in which von Helmholtz had his office. The officer tossed away his cigar, strode in and remained with the professor for more than an hour. It was Crown Prince Frederick, representing the German army and navy. Almost certainly he had come for advice on a military problem.

Students, on the other hand, rarely saw the inside of their professor's office. Once, for example, a research student from the United States—Michael Pupin by name—wanted to ask von Helmholtz a question: certain new developments in physics had not been discussed in the lectures he had been attending and he wondered why not. But when he told von Helmholtz's assistant that he wanted to question the great physicist, the assistant, Pupin said, "threw up his hands in holy horror." Such a question should not be asked; it would show a lack of respect.

As a teacher von Helmholtz was a disappointment; and so was the other renowned physics professor at the University of Berlin, Gustav Kirchhoff, known for his theoretical studies of radiation. His lectures were not badly prepared; quite the contrary. Planck described them: ". . . every phrase well balanced and in its proper place. Not a word too few, not one too many . . . dry and monotonous."

Despite their monotony and inaudibility, Planck had continued to attend these lectures, continued even when the audience of students dwindled to the point where there were only two besides himself. But what he learned in Berlin was the result of independent study. "I did my own reading," he said. Most of it concerned thermodynamics.

This part of physics is concerned with the relations between heat and mechanical action (work), and since heat is a factor in every physical system (in contrast to charge, for example), the range of thermodynamics is exceptionally broad. Not only the principles of engines may be deduced from its laws but also principles of weather, chemistry, geology, even the life sciences.

The fact that the laws of thermodynamics, simple and few, accounted for so much meant to Planck that they were truths, fundamental and absolute, expressing what in all of nature was simple, unchanging, eternal. He wished to devote his life to these laws, to explore their consequences in various realms of science, to demonstrate, in problem after problem, their endless applicability. Such logical demonstrations could, he believed, turn up fresh knowledge, for in his opinion it was not necessary to experiment in order to learn something new.

Therefore Planck began to study thermodynamics intensively, looking up the original literature on this subject, much of it the work of his two professors, Kirchhoff and von Helmholtz. But while he was being inspired by their past writings, they—like many other physicists—had lost interest in thermodynamics. The broad range of its deductions, the simplicity of its foundations, suggested to them that it was complete. If there was something new to be learned about the physical universe it could best be learned, they thought, from experimental results which did not fit theory, or from theory which was not so logically refined and coherent as that of thermodynamics. In short, while physicists, by and large, regarded the edifice of thermodynamics as beautiful but finished, Planck thought of its laws as a skeleton key, a key which could unlock endless doors and reveal the unknown.

In his solitary studies of thermodynamics Planck came upon an idea which, as he saw it, perfected that skeleton key. His recognition of this idea, which centered on something called "entropy," was a turning point in his life. Such a dominant part in his work did it play (this was the "hole boring" mentioned earlier) that one cannot tell about Planck without telling about entropy as well. While it is true that an understanding of entropy is not at all necessary in order to grasp what quantum theory means, it is essential for an understanding of how Planck arrived at that theory and of the sort of scientist he was. Therefore a few words about the laws of thermodynamics, for entropy is part and parcel of the second law.

The first law of thermodynamics states that energy always is conserved; it cannot be created from nothing and cannot be destroyed. This is a very general statement and today the first law is given much as it was when Planck studied it in his *Gymnasium*. The same is not true of the second law. It was discovered in connection with the study of engines, and in the beginning was intimately associated with the practical question, "How much work can be obtained from heat?" For while energy never is lost altogether, but changes into different forms, it is not possible to change all heat energy into work. In the natural course of events, said the second law, a certain amount of energy becomes unavailable for further use.

Today the second law is understood in a deeper and more general way, the idea "how-much-work" being only one of its many sides. Planck was one of the first to glimpse the deeper meaning when in his solitary studies at Berlin he happened upon the writings of Rudolf Clausius, who had worked out a version of the second law which differed from the accepted one.

To illustrate the difference between the two versions of the law, we can use two different statements which are less technical and compare them in terms of the amounts of information which they convey:

1. Lightning is seen before the corresponding thunder is heard.

2. The speed of light is about a million times the speed of sound.

The second statement contains the information given by the first. It is more general and at the same time it is more precise. It makes a universal statement and one that is measurable.

The version of the second law of thermodynamics due to Clausius resembles the second statement. It accounts for the observation that it is not possible to change all heat energy into work and for other observations as well. It accounts for them quantitatively. The measuring stick which makes this possible is called "entropy," and it is a purely mathematical quantity. Entropy is a ratio (a fixed relation between quantities) which measures nature's one-way changes, tending always to increase. "When a natural change occurs, entropy increases or, at best, remains the same." This was Clausius' version of the second law of thermodynamics.

Max Planck recognized this version of the law as superior. But its superiority was not evident to most other physicists at that time; it had not led to any startling discoveries and was, moreover, hard to understand. Entropy was a mathematical relationship; it had no connection with anything immediately tangible. Why should a law which appeared perfectly adequate as it stood be framed anew and in difficult entropy language? Why split hairs? This presumably, was the general feeling. In any event while indifference to thermodynamics was profound, indifference to the idea which "improved" it was even greater. With only one or two exceptions, "nobody at all had any interest whatever" in entropy when Planck happened upon it. This would change in the years to come. From the study of molecules, as we will see, a new meaning would be found for "the tendency of entropy to increase."

After his discovery of Clausius' work, Planck began his first scientific paper, which rephrased and refined Clausius' ideas. The paper is recognized now as a good, solid piece of work and he labored hard on it; enthusiastically, he "split hairs." When the paper was done, he submitted it as his doctoral thesis and it was published.

Later Planck described the effect of his paper on science. It was, he said, "nil." He had hoped that the men whose work had inspired his own would be favorably impressed, but Kirchhoff found fault with the paper and von Helmholtz, who probably had not bothered to read it, made no comment at all. Even Clausius remained silent. Planck sent the thesis to him and waited for a reply. When none came, he wrote again. This letter also went unanswered and so Planck made a trip to Bonn to see

Professor Clausius in person. The professor was "not at home" to the young man. The "nil effect" of Planck's work meant that his chance of moving up in the academic world was slight indeed. If his work was unknown, he never would be offered a professorship, and at this time a theoretical physicist had to teach in order to earn a living. He was not paid for research.

Not long after he received the doctor's degree, completing his formal education, Planck became *Privatdozent* at his home university, Munich. A *Privatdozent* was an apprentice professor who lectured without salary, receiving only a very small fee from students who chose to attend his lectures. From the ranks of these apprentices a few were chosen to become associate professors, the few whose lectures attracted students, whose research papers appeared regularly in the journals and found favor. Often a *Privatdozent* waited ten, fifteen years, even longer, before he won the longed-for professorship; many gave up and became *Gymnasium* teachers. Planck had an additional disadvantage: there were only a few universities which had a place for a theoretical physicist.

Five years later he was still *Privatdozent,* still hoping for recognition and advancement. He lived with his parents and yearned to be independent. It was an isolated existence: there was no one interested in his ideas with whom he could talk, and as before his letters to other physicists often went unanswered.

During these years Planck proceeded along the path which he had chosen in Berlin and all his life would follow: exploration of the laws of thermodynamics, and particularly the increase of entropy, measure of nature's one-way changes. In a series of papers he demonstrated that certain facts of physics and chemistry followed from a knowledge of entropy. These papers brought him no more recognition than his first had and for the same reasons. Someday, Planck believed, this situation would change. When the significance of Clausius' work was recognized then the value of his own also would be seen. He was correct in this belief except for one thing. He did not know that another theorist—an American, Josiah Willard Gibbs of Yale University—was following the same line of thought; and Gibbs's work had been published a little earlier than Planck's. Therefore when recognition of the entropy idea finally came it was Gibbs and not Planck who received the belated credit.

The years when Planck lived at home and worked as *Privatdozent* must have been a low point in his life, because he said later that when at last he was offered an associate professorship, at the University of Kiel, it "came as a message of deliverance." He was "supremely happy to accept," even though he suspected that it was not recognition of his ability which had prompted the offer but the fact that his father was a good friend of the physics professor at Kiel.

Not long after he moved to this city in northern Germany, Planck's work at last was read and appreciated by none other than Hermann von Helmholtz. This came about, strangely enough, as the result of a contest which Planck entered, and in a sense failed to win. Hoping to gain "somehow, a reputation in science," he had submitted a paper to the science faculty of the University of Göttingen, which was holding a competition. When the prizes were announced Planck learned that he had been awarded the second, that only two other papers had been submitted and that they had been awarded no prizes. Why, he wondered, had he not been given the first prize? He learned the answer to this question later, when the Göttingen faculty published the reasons for their decision: in one part of his paper, on the nature of energy, he had sided with von Helmholtz in a heated dispute the latter had been having with the professor of physics at Göttingen. It was this part of the paper which the Göttingen judges had singled out for criticism. They had withheld first prize, it appeared, out of loyalty to their fellow faculty member—

and by this act did Max Planck a great favor. For now von Helm-holtz learned of the obscure young physicist who had taken his side in the argument. He began to read Planck's publications, recognized their worth, and a few years later, almost certainly due to the influence of von Helmholtz, Planck was offered an academic plum, a professorship at the University of Berlin.

Planck was thirty-one years old when he joined the elderly, bushy-whiskered professors of Berlin. His figure was slight, his manner modest, and he had no whiskers, just a sparse, drooping mustache. Planck did not look like a professor extraordinary, as his title proclaimed (the full professor is called "ordinary" in Germany, his associate is "extra" to the "ordinary"). There is a story that once, not long after he came to Berlin, Planck forgot which room had been assigned to him for a lecture and stopped at the entrance office of the university to find out.

"Please tell me," he asked the elderly man in charge, "in which room does Professor Planck lecture today?"

The old man patted him on the shoulder. "Don't go there, young fellow," he said. "You are much too young to understand the lectures of our learned Professor Planck."

The young professor had not been warmly received by all his Berlin fellows. He was the only pure theorist on the faculty and some of the experimental scientists were suspicious of this young person who never entered a laboratory. (Einstein, later, would have the same experience at the same university.) But one friendship Planck made at the University of Berlin more than compensated him for the coolness of certain others. At last he was admitted to the intimate circle of *Exzellenz* von Helmholtz, at last he had someone to talk to who understood his ideas and would argue them out with him. From his exchanges with von Helmholtz, Planck learned more, he said, than he had learned during his entire formal education.

The young professor worshiped the older man. The few occasions when von Helmholtz praised him were, Planck said, "thrill-ing moments." And "when during a conversation he would look at me with those calm, searching, penetrating, and yet so benign eyes, I would be overwhelmed by a feeling of boundless filial trust and devotion. . . ." Yet for all his devotion to this fatherly figure and to the established laws of his science Max Planck soon would uncover a basic flaw in authoritative physics. Soon he would begin to work on the problem later known as "the ultra-violet catastrophe." First, however, something happened which gave him a tool he would use to solve this problem, something which demonstrated that Planck's allegiance to entropy had not been mistaken.

In Berlin, as in Munich and Kiel, Planck had continued his in-vestigations of the second law, as formulated by Clausius. Also, he had defended Clausius' version of the law in arguments with other physicists, many of these arguments conducted by letter. But although, with the increase in his academic status, his letters no longer went unanswered, all his attempts to persuade were futile. Planck was bitter about it. To him it was crystal clear that the entropy version of the law went deeper and said more, but others just could not see it. "Certain physicists actually regarded Clausius' reasoning as unnecessarily complicated and confused," he complains in his autobiography. "All my arguments fell on deaf ears." Planck felt bitter frustration of this sort more than once. He said in fact that not once in all of his scientific career did he ever have the satisfaction of convincing everyone that his ideas were correct. Always he had to wait until another person came along with quite different arguments to prove him right. In the case of the second law of thermodynamics this person, he said, was the Austrian physicist Ludwig Boltzmann.

Boltzmann was interested in the relationship between the laws of thermodynamics and molecular motion. At the time he was working on this problem, direct proof that there were such things as molecules had not been found. But if they did exist—and Boltzmann worked on this assumption—then their motion would constitute what is called "heat." Suppose one worked out what the motions of molecules might be under various conditions, according to Newton's motion laws. Perhaps, following this route, one would be led to the known laws of heat, the laws of thermodynamics, and find out more about them. For instance, one might learn why heat energy cannot be entirely converted into mechanical energy.

As an illustration, let us use the example of a penny dropped to the ground. When the coin comes to rest, heat is generated and this is equal in amount to the energy of motion which the penny has lost (its mechanical energy). Suppose now that the penny is heated until the exact amount of energy it lost has been restored. That will not cause the coin to move back to where it was in the first place. The coin will not move at all. Why? The reason becomes clear only when one considers the coin's molecular composition. Under all circumstances (Boltzmann assumed and it later was demonstrated), molecules are in motion and they move at differing speeds in differing directions. There is no sign of alignment, pattern, order. As the coin falls downward, however, all its molecules are sent at the same speed in the same direction. Although their tiny, random motions continue, there is an overall pattern in the motion, an order. That order is lost at the moment of impact. Adding heat to the coin does not restore the former orderly pattern; it merely speeds up the molecules' random motions. The particles do not align themselves so that they move in a single direction: therefore the penny does not move.

The reason why heat cannot be converted altogether into mechanical energy lies in the tendency of disorder to increase. "Natural processes move in one direction, toward an increase of disorder." Thus Boltzmann, on the basis of molecular motion, deduced the second law of thermodynamics. It was almost the same way that Clausius had given the law; his ratio, entropy, Boltzmann identified as a measure of disorder.

It was this work, Planck was sure, which won physicists over to Clausius' version of the second law. They were not persuaded at once; conclusive proof of the existence of molecules did not come until many years after Boltzmann's work. But when the physicists were persuaded, it was by Boltzmann's arguments rather than Planck's and it was to Boltzmann's conception of the second law, not Planck's. For in demonstrating a broader meaning of that law, Boltzmann also had demonstrated something else: *that it was not an absolute.*

Boltzmann's work was statistical. It was not possible then (or later) to measure the exact motion of each one of the countless molecules in a system (such as the penny, in our example). Therefore he had to base his work on assumptions about molecular motion on the average, statistical assumptions. The deductions which followed therefrom also were statistical: entropy increased, but only on the average. In a particular case, entropy might decrease. This was not impossible, only highly improbable.

"False," replied Planck. Just as energy was conserved in every single case, without exception, so did entropy increase, or remain the same, but *never* decrease. It was not a matter of probabilities, of varying shades of gray, but of pitch black versus pure white. By letter, Planck argued this difference of opinion with Boltzmann, who had a biting wit and did not hesitate to use it. The argument, at least on Boltzmann's side, was not a friendly one.

Half a century later, when he wrote his autobiography, Planck's wounds from the battle over the second law still were smarting. "I was not to have at all," he said, "the satisfaction of seeing myself vindicated." All his attempts to persuade had been, he felt, just so much wasted effort and wasted time. He might as well have kept silent; Boltzmann still would have come along to settle the matter with his repugnant statistical ideas. Yet in spite of this rankling argument, in spite of his dedication to entropy increase taken in an absolute sense, Planck at last accepted Boltzmann's statistical interpretation of entropy and himself demonstrated its soundness.

In the next chapter we will see how this happened, as Max Planck, a most conservative person, introduces what has been called "the most revolutionary idea which ever has shaken physics"—the quantum theory.

CHAPTER FOUR

Max Planck: The Quantum Theory

I was not to have at all the satisfaction of seeing myself vindicated.

—Max Planck

A YOUNG PHYSICIST who was staying in Max Planck's home during a visit to Berlin gradually became aware of a certain regularity in the Planck household. The visitor's curiosity was aroused; he made an observation: stationing himself by the door of his room he waited to see what would happen as the clock in the hall struck the hour. Sure enough, while the clock still was sounding, Planck emerged from his room, proceeded down the stairs and then out the front door. Further observation confirmed the regularity of his comings and goings; as the big hall clock struck a certain hour, invariably there Max Planck would be, on his way down the stairs.

Systematically, part of Planck's day was allotted for a walk, just as thirty minutes regularly were devoted to playing the piano. And when this highly organized scientist worked in the study among his treasured collection of scientific volumes, he worked standing up. His desk was a high one like those used in Dickens' time by clerks, who sat on high stools. But Planck did not use a stool.

It was in 1897, nine years after he joined von Helmholtz in Berlin, that Planck began to work, hard and upright, on the problem later called "the ultraviolet catastrophe," the problem which

51

led him to the idea that energy is not continuous, contrary to prior observations and to the science based on these.

As mentioned earlier, Planck was not the only person to become interested in the problem of accounting, with physical theory, for the light and heat sent out by the illumination standard called a "black body." Physicists were attracted to this ideal case where the radiation spectrum, uninfluenced by other factors, depended only on temperature; others besides Planck were trying to account for the ideal case and thus gain an understanding of the radiation process in all cases.

Radiation similar to that emitted by a black body may be observed when any solid piece of metal, such as an iron poker, is heated. At low temperatures long-wave radiation is emitted, that of the infrared region of the spectrum. As the temperature rises, radiation of shorter and shorter wavelengths also appears, and the poker glows red, then orange until with the addition of other colors it appears white to the eye. A further increase in temperature will produce even shorter wavelengths which are not visible, those of the ultraviolet end of the spectrum.

The spectrum of the black body (or of anything else) shows how energy is distributed among the different wavelengths, for some colors are brighter than others. In Planck's day it was already possible to measure in the laboratory the energies represented in a spectrum. Those of the black body, at various temperatures, were known by experiment. The problem was to account for this specific energy distribution on the basis of general knowledge. Broadly speaking, this is the way physicists attacked the problem: they began with a hypothesis as to the cause of radiation, a hypothesis which seemed reasonable on the basis of what already was known. This hypothesis took the form of a model illustrating how radiation might arise from matter. Then physicists deduced the radiation energies which would occur according to the model, checking these theoretical deductions against laboratory measurements of the actual energy spectrum, hoping to find confirmation of the hypothesis.

The models which were used in attempts to solve the black-body problem were not models of the atom. Evidence was just being found at this time for the existence of the electron, and speculations as to atomic structure lay in the future. Physicists worked with rough general assumptions as to the material structure responsible for radiation. They postulated that the black body was composed fundamentally of some sort of electrically charged particle, the motion of which—speeded up by heat—produced the radiation. Actually such a rough hypothesis was adequate for the problem at hand; these were not the assumptions which were responsible for the failure to account for black-body radiation. The trouble came when physicists took the next step: in order to arrive at the distribution of energy in the spectrum, one first had to assign energies to the particles responsible for the radiation; energy had to be distributed among them so that motion of such-and-such kind yielded radiation of such-and-such energy. But if energy is continuous, as was assumed, there can be no restrictions on it, for these would introduce jumps, abrupt change. Therefore the oscillating motion of the electrical particles could not be restricted; the oscillations had to be indefinitely small. But from this it follows inescapably that the radiation energy at the short-wave (high-frequency) end of the spectrum will be indefinitely large. Speaking more technically, there will be a continuous rise to infinity with declining wavelength. Thus physics predicted ultraviolet catastrophe, for we know that a glowing material body does not radiate an infinite amount of energy. Moreover, experiment shows most of the energy appearing as waves of medium length. The solution is obvious: to avoid the ultraviolet catastrophe and reach an accurate description of black-body radiation one must limit the possible energy in such a way that it is not infinite and does not go chiefly to the shortest waves.

This is easy enough to see now—afterward. When Planck worked on the problem it was far more tangled. For one thing, experimental results were not as clear-cut then as they became later; one would seem to be on the right track only to be proved wrong by some new and more accurate measurements of the radiation. What is more, it was virtually impossible at that time to recognize the assumption of continuous energy as the root of the trouble. Black-body measurements were the first recognizable evidence that this assumption might be wrong; there was no background of unsolved problems to make one suspicious. Indeed, the idea of continuity in nature, and therefore unbroken energy, did not appear to *be* an assumption. It was taken for granted. If, for example, one were trying to work out laws governing the growth rate of a plant variety, one would take certain things for granted, such as the fact that the sun will rise every day.

That was the way physicists "assumed" that energy was continuous. The fact that they were unable to solve the problem on their assumption did not lead them to question it. They had made numerous others as well, and it was much more reasonable to question these. Only much later did they understand that the laws and theories they were using rested on a questionable idea and that whether one treated the problem in terms of heat, mechanics, or electricity, one must come to the same conclusion, the ultraviolet catastrophe. All routes led to it.°

And Max Planck continued to trust the physics of his time even as he solved the black-body problem with an idea which contradicted that physics. That this was possible was due chiefly to two things: in the first place, Planck solved the problem back-

° Einstein was the first to point this out, five years after Planck's work.

ward, so to speak; he put together a formula which correctly described the energies of black-body radiation without understanding the full implications of the mathematical expression he had written. Secondly, in finding the meaning of his formula, he used a mathematical procedure which was new to him and which, without realizing it at the time, he misapplied. Let us see how this happened.

Planck arrived at his first radiation formula by doing some juggling. Among the different formulas which had been worked out in connection with the problem, there were two which were partly sound: one gave the energy distribution of the short-wave region of the spectrum correctly; the other accounted exactly for the long-wave region. Planck saw a simple, logical way to combine the best features of the two mathematical expressions. Whether this new formula represented the exact distribution of energy throughout the entire spectrum at every temperature he did not as yet know. But he did not have to wait very long to find out.

His formula was announced almost at once at a meeting of the Berlin Physical Society on October 19, 1900. Among those present was Heinrich Rubens, who had been working on black-body experiments. As soon as the meeting was over, Rubens went to his home and stayed up most of the night, comparing his measurements with Planck's formula. The agreement was excellent. Rubens appeared at Planck's home the next morning to tell him the good news.

Now Max Planck knew that he had found the solution to the black-body problem, but neither he nor anyone else knew what that solution meant. He had gotten it by guessing, by putting parts of formulas together, and in doing so he had automatically changed the assumptions on which these mathematical summations had been based. How had they been changed? What new assumptions had he, without knowing it, introduced? The only way to find out was to try different hypotheses and see if one of them led to his formula. Only then would he understand what the successful combination of symbols was saying about the process of radiation.

Planck began his attempt to draw a logical line from theory to fact so that fact not only was described but understood. For two months he tried to do this, working harder, he said later, than he ever had worked before or would again. Although the problem as we have defined it concerned electrical particles and their motion, Planck did not base his attempt to solve it on electrical theory. Instead, he used thermodynamics. As mentioned earlier, the range of this science is very broad and the black-body problem could be expressed in the vocabulary of thermodynamics as well as in the vocabulary of electrical theory.

Planck had a reason for using thermodynamics. Like other physicists, he wanted to solve the black-body problem, but he wanted to do something else as well: he wanted to demonstrate, by solving it, the fundamental nature of the second law of thermodynamics. He wanted to show, once again, the increase of entropy taken in an absolute sense. That is why he had chosen the black-body problem in 1897. That is why for four years he had been trying to solve it. He believed that if he could demonstrate the increase of entropy in the case of black-body radiation this would lead automatically to the correct solution to the problem. For four years he had been unsuccessful. When, in 1900, he found the correct solution, but not on the basis of theory, he returned at once to his earlier work and tried to deduce that right answer in terms of the increase of entropy, the increase of entropy in every single case.

It could not be done. This meant that either he must give up his four-year attempt to solve the problem on the basis of the second law of thermodynamics or try one last possibility: Boltzmann's statistical interpretation of that law. Planck never had used this interpretation. Suppose he turned to it now and defined the problem anew, as Boltzmann would, in terms of molecular motions and their statistical probabilities, and then, on *this* basis, was able to deduce his own correct formula? Automatically he also would be demonstrating that the second law was not an absolute, black-and-white certainty.

Thus Max Planck's systematic, patient, disciplined pursuit of this law brought him to the point where if he were to continue he might have to deny that it was an absolute truth. He did go on: he looked up the papers in which Boltzmann had presented his statistical method, applied it to the problem at hand and found that, indeed, he was led to his own formula. In this way he demonstrated that the increase of entropy, while overwhelmingly probable, was not absolutely certain, as for several decades he had conceived it to be.

If this were all Max Planck had done in physics, it would have been a great deal. But at the same time he did it he also stumbled on a new idea. In order to deduce the correct black-body formula it was necessary to follow a statistical route, but something else also was necessary: a hypothesis which broke drastically with nineteenth-century physics. By not applying Boltzmann's method as Boltzmann had intended it to be applied, Planck in effect based his work on the hypothesis that energy is not continuous. Let us see how this happened.

At a certain stage in solving a problem, Boltzmann's method required one to treat energy as if it were divided into separate portions, because in order to determine a degree of probability it is necessary to have something to count up. Boltzmann and others familiar with his method understood that this dividing up of the energy was merely a calculating technique. At a later stage in the calculations one always, by means of another technique, got rid of the portions, making the energy continuous again.

A like procedure sometimes is followed in solving problems which involve a curved figure such as, for example, a circle. Rather than calculating the exact circumference of the circle, it often is easier in terms of arithmetic to use a rough estimate instead. In effect, the mathematician gets rid of the unbroken curved line, which is difficult to handle arithmetically, and represents the circle as closely as possible by means of tiny straight lines equal to each other in length. (Such a figure if drawn would look almost exactly like a circle.) The total length of these straight sides may be taken for purposes of calculation as the circumference. Then later on, when it is more convenient, the mathematician obtains the exact circumference figure by letting the number of straight sides increase without limit. In this way he restores the curved unbroken line.

An unbroken flow of energy, like a curved line, is infinitely divisible. Had Planck applied Boltzmann's method as its originator had intended and, after calculating with portions of energy, put the pieces together, allowing them to increase without limit, an entirely different conclusion would have been reached. Assuming that his calculations were correct in other respects, he would have obtained the formula containing the ultraviolet catastrophe. As mentioned earlier, one had to come to this regardless of the route taken. During all the years Planck had worked on the blackbody problem, he had chanced to avoid it; his methods never indicated the disastrous consequence of classical physics. When, finally, he came to the crucial step, he was using a procedure which was new to him. He did not join up the energy portions. At just that point he recognized that there was a way to reach the right answer: the formula he had before him. This meant that he avoided the ultraviolet catastrophe. For energy treated in portions is not infinitely divisible and therefore the radiation energy also is not infinite in quantity. What is more, by making the portions unequal, energy can be distributed in such a way that it does not go chiefly to the waves of shortest length.

That was how Planck found the simple but strange rule which lay behind his first formula and which forms the basis of the quantum theory. The rule states a relationship between a portion of energy, which Planck named a "quantum," using the Latin word for "how much," and the frequency of a wave, which is related to its length. To find the energy of a quantum E, one multiplies a wave frequency f by a fixed number, a constant, represented by h, or

$$E = hf$$

Planck's rule was strange because it announced an equality between energy conceived as discontinuous (E) and (since frequency pertains to wavelike phenomena) energy conceived as continuous. The full impact of this equation would not be felt until later, when it was shown that light could be understood as a stream of energy grains, or quanta. The black-body problem concerned the emission and absorption of light, not the structure of the radiation itself. Planck's work set a limit only on the energy of the electrical particles which compose matter: they might move in certain ways only so that a whole quantum hf was emitted or in turn absorbed. If light comes out of matter in portions and goes back into matter in portions, it would seem to follow that light must exist in particlelike form. But in 1900 there was no evidence whatsoever that light might have such a structure while its wave properties had been established beyond question. The broader extension of $E = hf$ would come later when there was some evidence, just a little, for the discontinuous structure of light. Until then the discontinuities announced by Planck's equation were ascribed only to the emitter, not to what was emitted.

Planck's work did not directly challenge the wave theory of light; still it broke drastically with accepted ideas, ideas based on the observation of nature. A swinging pendulum appears to lose its energy gradually and smoothly and here the pendulum stands for almost anything which is large enough to be observed directly. In every case, energy changes appear to be continuous.

Yet Planck's work denies this: in the case of black-body radiation the energy changes in bursts or jumps. The jumps are exceedingly small. In the equation $E = hf$, the constant h indicates their size and (in terms of the unit erg-seconds) h stands for the number

$$.000\,000\,000\,000\,000\,000\,000\,000\,006\,6$$

or (6.6×10^{-27}).

Then was Planck's work so revolutionary after all? The jumps simply were too small to be detected by ordinary means of observation. Up to about 1900, physicists had been concerned mainly with objects and happenings on the pendulum scale, or larger. Now they were beginning to probe more deeply and had found evidence that on a smaller scale nature did not conform to expectations. But physics still could be used where it had been used before. One would not have to consider minute discontinuities in energy in order to understand the motion of the pendulum, or the planet. Planck, in showing something new, had not necessarily demolished the old—but he had shown that there were limitations to it.

And it was just that which made his work revolutionary. It suggested that the understanding which classical physics represents might not after all be the last word. It hinted that the physical world might just possibly not be a giant mechanism whose workings are in an ultimate sense foreseeable. Planck had planted a seed of doubt. From this very gradually a new physics would grow as more was learned about the atomic region of nature. From this beginning would come an understanding of why statistical rules, as in the case of radioactive change, must be used; of why the atom does not ordinarily glow with light and why its electrons do not bury themselves in the nucleus.

Two weeks before Christmas, in the year 1900, Max Planck announced the results of his theoretical work on black-body radiation. In his speech to the Berlin Physical Society he said nothing of the challenge to classical physics which this work represents, and other evidence strongly suggests that he did not at first recognize the full implications of his mathematics. He did not know that the route which led to $E = hf$ was not Boltzmann's route.[*]

What is more, when Planck later on did recognize the revolutionary character of his work he tried to undo it. Returning to the problem which he already had solved, he tried to find a different solution, a different theoretical route which also led to the right answer and did not require the fatal equation $E = hf$. For years he tried, to no avail. But he did not give up hope. Perhaps future discoveries still would make it possible to mend the break with classical physics which he had made. Planck continued to hope for this even though it would diminish the importance of his own work.

Instead, the years brought more and more confirmations of this work, as well as more and more honors in recognition of it. In Germany, Planck's position would be that of a Hermann von Helmholtz. Everywhere in the world physicists would regard him as the father of a revolution in their science. Again, Max Planck did not have the satisfaction of seeing himself vindicated.

In the years after 1900, much of his research was devoted, as before, to the second law of thermodynamics. Now he used Boltzmann's statistical interpretation of that law but, as before, he studied the consequences of the law in case after case, problem after problem. Rarely, if at all, was the outcome of these

[*] It is not possible to know for certain what Planck did or did not recognize at that time. This account relies on the opinion of a historian of physics, Martin J. Klein, who has made a detailed study of the line of reasoning Planck followed in his work on black-body radiation.

labors dramatic. Some of Planck's colleagues wondered why he found such "thankless questions" worth pursuing. They thought he ground away like a machine, the victim of his own great energy, disciplined habits and precise, tidy methods.

Einstein disagreed. After 1913 he came to know Planck quite well and in spite of their differences in attitude and background, and in spite of Planck's formality which often made Einstein uncomfortable in his presence, he recognized something in the older man that united them. Planck did not work like a mechanical man, said Einstein. It was a longing to find harmony and order in nature which spurred him on, "a hunger of the soul." That is why Einstein, speaking to other physicists, referred to "our Planck." "And that," said Einstein, "is why we love him."

It was Albert Einstein's work which caused many physicists to recognize the importance of Planck's. At first the quantum idea was ignored. For five years after 1900, physicists did not take it up; they did not look for other cases where the quantum hypothesis might account for observations. Why not? Historians of science have found several reasons. One of them is the burst of discoveries made just at the turn of the century—in 1895, the discovery of X-rays; in 1896, radioactivity; in 1897, the electron; in 1898, radium—discoveries which made it easy to overlook Planck's work of 1900.

Another reason for the indifference is to be found in the nature of Planck's work, which represented a challenge to physics as it was then constructed and understood. Physicists do not necessarily take up such challenges at once. They tend to wait until the challenger's case is overwhelmingly strong. In physics, theories come and go. Often what appears sound, later must be abandoned in the light of new information. Thus for all one knew in 1900, the quantum theory might be just a clever way of getting around partial information. It might not stand the test of time.

What is more, the physics challenged in the single instance of black-body radiation otherwise worked well. It solved problems; it enabled one to find things out. Few physicists are inclined to look for faults in their idea tools when these tools are useful.

Given this situation, one may make a prediction as to the kind of physicist who would be likely to take up the challenge which Planck's work represented. He would be interested not in the application of physical laws but in the structure of ideas behind these laws. He would be critical of this structure, willing to entertain ideas which run counter to it. He would want to go to the root of things.

Albert Einstein was just such a scientist. In the next chapter, we will see him at work on the two major theories of contemporary physics as he takes one, the quantum theory, a great step forward and introduces the other, relativity.

CHAPTER FIVE

Albert Einstein:
Work of 1905

One is never again as intelligent as one is at sixteen.
—Leo Szilard

IN THE FIRST YEAR of the twentieth century, when Max Planck's work on black-body radiation appeared in a scientific journal, Albert Einstein was twenty-one years old, a graduate of a polytechnic institute and unemployed. He had applied for a position as *Privatdozent* at the Swiss institute where he had studied and had been told that there was no opening. Then he had tried to obtain a position teaching at a *Gymnasium*, but without success. At length, he had found an advertisement in the newspaper for a private tutor. The stocky young man with abundant curly hair and sad-looking dark eyes answered the ad and obtained the position.

His pupils were two boys who were not doing well in school. Einstein had been a poor *Gymnasium* student himself. He had objected strongly to what he called the "education machine," which, as he saw it, force-fed knowledge so that it might be disgorged at examination time. "Curiosity," he said, was a "delicate little plant [which], aside from stimulation, stands mainly in need of freedom." Now he taught his two students as he wished that he had been taught. Instead of feeding them solutions to be committed to memory he asked them questions and encouraged them to find the solutions for themselves.

He liked the work, but one thing was wrong. The boys still were attending classes at the *Gymnasium,* where Einstein felt that their curiosity, which he was doing his best to stimulate, was daily being smothered. He went to his employer and asked him to take the boys out of school, explaining that he could teach them better than could their *Gymnasium* teachers. This suggestion was not well received, which is perhaps not surprising in view of the fact that Einstein's employer himself was a teacher, a teacher in a *Gymnasium.* The new tutor was fired.

Again Einstein was without work and again he could find none, until a friend introduced him to the director of the Swiss patent office, in Berne. After giving Einstein a long written examination, the director decided that the young man, although inexperienced, might be useful in his office and hired him.

Einstein liked this job too. He was given a pile of patent applications in which inventors described their ideas, described them in great detail, as a rule, and with unfamiliar technical terms. It was his job to pick out the invention's essential features and then write a new simplified description so that his superiors might decide whether the invention was eligible for a patent. Einstein liked to use his thinking abilities in this way. He was interested in technical apparatus and instruments and occasionally would come upon an ingenious idea in his daily pile of letters. That made the job rather exciting. Best of all, he found that he could complete a day's work in three or four hours. Then he would begin to think about his own work, physics, and about some problems which he was trying to solve. His calculations were done on bits of paper which he would thrust hastily into a drawer when someone came into his office.

One problem on which he worked has been mentioned earlier in this book, in connection with Ludwig Boltzmann and entropy: the problem of formulating thermodynamic laws in terms of the probabilities of molecular motion of various kinds. Einstein, with a specific purpose in mind, developed and extended Boltzmann's work. This involved working out a mathematical procedure which enabled him to state the overall motion of a system in terms of the laws of chance—that is, a statistical mechanics. Einstein did not know that Josiah Willard Gibbs—the same Gibbs who had published his findings on entropy earlier than Planck —already had formulated such a mechanics. Once again, different men, working in ignorance of one another, had been attracted to the same problem at about the same time.

The fact that most of his work had been done earlier by another meant little to Einstein when he learned of it. He had forged his statistical mechanics, his statistical thermodynamics, as tools. Now he used them. He wanted to convince the many scientists who still doubted that there were such things as molecules (and therefore atoms). Using his tools, he calculated that under certain conditions motions due to molecules could be observed by microscope. If particles of a certain mass and size were suspended in a fluid, their movements would reflect their collisions with the molecules of which the fluid was composed. The average motion of one would be the same as that of the other. Thus Einstein made a prediction on the basis of the hypothesis that molecules exist, the hypothesis on which his statistical laws, like Boltzmann's, rested. Now an experiment could be done to see if the prediction was correct.

As a matter of fact this experiment already had been performed (which was something else Einstein learned only later). An English botanist, Robert Brown, had noted the constant zig-zag motion of tiny particles of pollen suspended in a fluid, a motion which was not caused by any external influence. He had noted it seventy-eight years earlier and the motion had been named after him, "Brownian." Now Einstein accounted for that motion on the basis of the molecular hypothesis. Further observations of Brownian motion showed that Einstein's predictions were correct in detail: the particles moved in just the way they would if there were such things as molecules. This constituted the first visible evidence for the existence of molecules and did indeed convince many doubters among the community of scientists. At the same time it demonstrated the importance of the mathematical tools prepared in advance by Boltzmann, Gibbs, Einstein, and others.

Brownian motion was one problem which Einstein worked out at the patent office. The theory of the photoelectric effect was another and it was for this that he won, in 1921, the Nobel prize.

The photoelectric theory is concerned with the structure of light. For some time Einstein had been thinking about certain ideas at the root of physics which ran counter to one another. On the one hand was matter: composed of discrete particles, atomic, discontinuous. On the other was radiation: insubstantial, wave-like, continuous. Was this the whole story, he wondered, or was there perhaps an underlying unity? Might not radiation also have a discontinuous aspect, at least under *some* circumstances?

There could be little doubt that the structure of radiation, when it was traveling through space, was wavelike (and continuous, for the two go together). A beam of light, for example, may be split in such a way that light added to light gives darkness. Only a wave picture can explain this behavior: the crest of one wave, coinciding with the trough of another, cancels out light. Under *these* circumstances, light must be a continuous phenomenon. But what of other circumstances? Was the case for continuity airtight?

It was against this background that Einstein viewed the problem of black-body radiation. He knew the work Planck had done on this problem several years earlier, but found it confused. Using his own statistical tools, he explored the question, "What conclusion follows logically from physics when applied to the black-body case?" and satisfied himself that the conclusion, re-gardless of what route one took, had to be the ultraviolet catas-trophe. Here the assumption of continuous energy broke down; here was some support for the idea that the case for continuity might not be airtight. While Max Planck had looked at the prob-lem of black-body radiation as a puzzle to be answered on the basis of what was known, Einstein recognized it as experimental evidence of something new. His calculations, like Planck's, led to the conclusion that light was emitted from a black body in por-tions and absorbed in portions. Very well. Perhaps under such circumstances light always behaved as if it were composed of portions. Other experiments had been done on the emission and absorption of light. What had they shown? Were their re-sults in accord with the continuous wave picture of light? Or with the quantum hypothesis?

This line of thought brought Einstein to the work of a German experimenter, Philipp Lenard, who had studied the absorption of high-frequency light (such as ultraviolet) by certain metals. Under these conditions so much light energy is transferred to electrons in the metal that some of them are bounced out of the metal surface. This bouncing out of electrons by high-frequency light is called the "photoelectric effect."

In his experiments Lenard had used beams of monochromatic light (a single color, a single frequency), training the beam on a metal strip and then measuring the energies of electrons which were ripped out of the metal. When the light source was moved closer to the metal target, the intensity of light was increased and therefore its energy. Under these conditions, the electrons which were ripped out of the metal also should have had greater energy. They should have moved faster. Lenard found that they did not. An increase in light intensity meant that *more* electrons were released but their speed was no greater.

Here was a case where the continuous-wave picture of light

failed to account for observations. Suppose one changed the picture. Suppose one assumed that a metal strip, when struck by light, was being bombarded by a shower of energy quanta—"photons," as they were later named. Striking the metal, a photon of light would transfer its energy to an electron, either sending it more deeply into the metal or ripping it out. Each photon, each quantum, could have but one definite amount of energy, no more no less. When the light source was moved closer to the target, more photons would strike the surface, but the energy of each photon would not have changed. Therefore the speed of the ejected electron also would not change. An increase in light intensity would mean an increase in the number of photons striking metal. Due to the greater number of collisions, the number of electrons released also would be greater.

While the wave picture of light failed in the case of Lenard's results, the quantum picture fit perfectly. As we have seen, Planck's solution of the black-body problem rested on the equation $E = hf$, making the energy of a quantum dependent upon the frequency of a corresponding wave. If the same equation held also for the photoelectric effect, Lenard, as he experimented with light beams of different frequency, should have observed a difference in his results. The photon associated with ultraviolet light would possess more energy than the photon of red light and therefore the electrons it ripped out of metal would travel at a faster speed.

Lenard *had* found this to be the case: the speed of the ejected particles differed with the frequency of light he used, and as Einstein showed, the speeds differed in accordance with the equation $E = hf$. The quantum hypothesis accounted for observations of the photoelectric effect just as it had accounted for the energies of black-body radiation.

This was the way Einstein extended Planck's quantum idea to widen the rift with classical physics and underline the contradiction stated by $E = hf$, with its energy particle on one side and wave definition (f) on the other, separated by an equal mark, an equality between opposites. Twenty years would pass before physicists would understand how light could have properties which are particlelike and also wavelike. In a later chapter, we will come to this.

Einstein's paper on the photoelectric effect was published in a German scientific journal in October, 1905. In the same issue his paper on Brownian motion also appeared. In the same issue, there was a third paper, with the same signature: Einstein's first theory of relativity. All of this work was completed when he was twenty-six years old. All of it was done while he was a clerk in the Swiss patent office, done in the "free" time when he worked with one eye on the door, so to speak. And it was done at home in the evening, after working hours. These might not seem ideal conditions for thinking, but Einstein appeared to find them so. These were his most productive years; quite possibly, they were also his happiest. His formal education, which he had thoroughly disliked, was behind him. He had left his native country, Germany, which he had found gloomy and oppressive. He was financially independent and was free much of the time to pursue his own work. All of his life this particular freedom meant more to him than anything else.

Even as a small child, Einstein was calm, quiet, thoughtful. He learned to talk much later than is customary, to the worry of his parents, who feared for a while that he might be mentally retarded.

The Einsteins had come to Munich from a small town—Ulm, in south Germany—a year after their only son was born. (There was one other child, a daughter.) In Munich the father, Hermann Einstein, started a small electrochemical business. It was not very successful but earned him a moderate living, at least for a while. He and his wife have been described as cheerful, easygoing, small-town people. They were not intellectuals but felt a profound respect for education and wanted their son to do well in school.

But his worst subjects were the two which were generally considered the most important, Latin and Greek. While he read many books, rarely were these his school assignments. Much of the time he did nothing. He was daydreaming.

In the short autobiography which Einstein wrote when he was in his sixties, and which he jokingly called his "obituary," he

Albert Einstein in the patent office at Berne, Switzerland. When this photograph was taken he was about twenty-four years old and working toward the conclusions published in 1905.

recalled some of the events which inspired his daydreams. One such event was his first acquaintance with a compass, when he was four or five years old. He saw that no matter how he shook and turned the compass its needle pointed the same way. His own actions did not control it; the needle answered the command of something which he could not see or feel, something hidden. There was a world beyond the one of his immediate perception, an unknown world. He felt "wonder" he said; indeed the experience made him tremble and turn cold.

When he was twelve years old there was another event, one which marked a turning point in his life. Knowing that in school he soon would be studying plane geometry, he happened to glance through the geometry text to see what was in store for him. A "subject" could be quite interesting, he had found, *before* school obligated one to study it. In this case, he found the textbook much more than interesting. He was profoundly impressed by the logically ordered proofs for each assertion, by the close connection between diagram and reasoning. Here he saw certainty, order, beauty. Like Max Planck, Einstein was inspired by the recognition that it was possible to find a meaningful pattern in the universe. But unlike Planck, Einstein did not recognize this in the classroom.

It was the same in the case of music. Like Planck, Einstein's love of music was second only to his love of physics but, again, he had to discover music for himself. At his parents' insistence he took violin lessons, starting at the age of six, but remained indifferent to music until, when he was thirteen, he heard Mozart's sonatas and, by himself, began trying to play them on his instrument.

As we have said, the discovery of plane geometry when he was twelve marked a turning point. Until then, Einstein had been strongly attracted to religion. His parents, descendants of Jews, were not at all religious themselves. Instead of sending their son to a Jewish elementary school as was customary, they enrolled him in the school closest to their home. It happened to be a Catholic school. There Einstein absorbed the ideas and ritual of that faith and became ardently religious, while at home his father made fun of him for such beliefs.

These were abandoned abruptly, however, as inspired by the geometry book, Einstein began to study mathematics and to read accounts of science for the layman. Convinced that much in the stories of the Bible could not be true, he turned against organized religion with feelings as strong as those which had drawn him to it. For a while he was, he said, "an anti-religious fanatic," seeing the church as an authority, like school and like the army. It seemed to him that all these institutions were the same, a single "education machine" into which the individual was fed as a child and trained to think and believe in one way.

Kadavergehorsamkeit, the obedience of a corpse—this was required of soldiers in the imperial Prussian army, the army of the German nation. Einstein felt that corpselike obedience was the theme at school as well. He saw the students as privates, standing at attention, reciting at command. He preferred punishment to reciting something which he could recall but did not understand. As far as possible he did not join in. All his life he remained a nonjoiner; and although with time he came to judge human institutions a little less harshly than he did as a boy, he remained skeptical of "the convictions which were alive in any specific social environment."

Others might have found such isolation lonely; Einstein did not. He wished to free himself as much as possible from what he called "the chains of the merely personal, from an existence which is dominated by wishes, hopes, and primitive feelings." He had felt this desire before he was twelve years old; indeed he said that that was why he had been so attracted to religion. Between Catholic doctrine and Jewish traditions he had seen a resemblance. Different symbols were used by the different religious systems to express what was fundamentally the same. Underlying apparent differences was order, an order in which the "merely personal" part of life could be submerged.

He had rejected this "religious paradise of youth," as he called it, and had substituted something else in its place: the beauties of geometry, the realization that understanding of the physical world was possible. "Out yonder," he said, "there was this huge world, which exists independently of us human beings and which stands before us like a great, eternal riddle. . . . Contemplation of this world beckoned like a liberation. . . ." In this way also one could escape the merely personal chains.

Thus at about the age of twelve Einstein entered upon a road which he would follow for the remainder of his life, a road which led away from other people and their institutions and which was not "as comfortable and alluring as the road to the religious paradise." Yet this route also, he believed, led to a kind of paradise and he used the language of religion when he spoke of it. "God," he would say, "is sophisticated, but He is not malicious"; and by this Einstein meant that the universe is made in such a way that man *can* comprehend it, hard though this task may be. Again he might say, criticizing a theory of physics as not touching upon fundamentals: "The theory yields much, but it hardly brings us nearer to the secret of the Old One."

When Einstein was fifteen years old, his father's business failed. Hermann Einstein decided to give it up, move to Italy and start over again; but his son Albert was not to go with the rest of the family. He must remain in Germany until he completed his studies at the *Gymnasium*. Without a diploma, he could not enter a university and without a university education most intellectual professions would be closed to him. Einstein therefore remained in Munich, struggling with Greek and Latin,

at odds with dictatorial schoolmasters and submissive school-
mates. What he knew of Italy he loved: its sculpture and painting, its music. There it would be warmer—and gayer.

In a boardinghouse Einstein pondered a practical question:
how, without compromising the future, escape from school to
Italy? After obtaining two documents he thought he had solved
it. One from his mathematics teacher stated that Albert Einstein's knowledge of mathematics was such that he was qualified
to enter a more advanced school. With this Einstein hoped to
gain admission to a technical institute outside Germany even
though he did not have a diploma. The other document, which
he obtained from a doctor, stated that due to nervous exhaustion
Albert Einstein must join his parents in Italy for a six months'
rest. But before he presented this to the school authorities they
solved his problem for him in another way by asking him to leave
school. When he asked the reason, he was told that others in his
class had been influenced by his negative attitude and had shown
their teachers disrespect.

Soon after his arrival in Italy, Einstein renounced his German
citizenship and having cut this bond also severed his formal connection with the Jewish religious community. But there was to
be no escape from school.

After a brief and rapturous exploration of Milan, his parents'
new home, and a hike over the Apennine mountains to Genoa,
his "rest" was cut short when his father said that he could support him no longer. The business was no more successful in Italy
than it had been in Germany; his son must complete his education (a relative had agreed to pay for it) and then find a job.

Einstein had hoped to enter a polytechnic institute where his
knowledge of mathematics might make up for his lack of a diploma. One of the best of these institutes was located in Zurich,
Switzerland, and he applied for admission. But the statement
which he had obtained from his mathematics teacher did not

gain him entry as he had hoped. It was necessary to pass an entrance examination. Einstein failed it. There was nothing for it now but to go back to school and get his diploma. This he did, in Switzerland, and a year later when he again applied to the Zurich Polytechnic Institute he was admitted.

He studied physics. Mathematics attracted him but there were many branches of it and mastery of any one, he felt, would take a lifetime. He intended to use mathematics as a tool with which to express fundamental physical laws and try to "unriddle" the universe, but which branch of mathematics would prove to be the proper tool? He must choose—and he found that he could not.

This was not the case with physics. Although that science also was divided into many separate fields of study, each one difficult, he was able, he said, "to scent out that which was able to lead to fundamentals and to turn aside from everything else." (Later, however, he would regret that he had not studied more mathematics. For many years his second theory of relativity was held up as he searched for the proper mathematical tool.)

Turning aside from everything else—and this included most of the subject matter he had to know if he were to pass the final examination—Einstein studied physics in his own way. Availing himself of the freedom *not* to attend lectures, he spent long hours in the physics laboratory. Instead of studying the texts assigned, he read books about physics which he chose for himself. Like Max Planck, he was interested in the writings of von Helmholtz and Kirchhoff, which gave the logical background for accepted theory. From these works and others like them he learned about the structure of physical theory. At the same time, he was trying to solve a problem: he was studying the accepted theories of physics critically, in search of a clue—scenting out "that which was able to lead to fundamentals." Soon we will learn what the object of his search was.

The Zurich Institute offered him more freedom as a student than he had known before but there was still, as he said, "the hitch . . . the fact that one had to cram all this stuff into one's mind for the examinations, whether one liked it or not." The time for examinations drew near and he was not prepared. He had a friend, however, who conscientiously attended lectures, who had taken notes which were well organized and complete. Einstein borrowed and used them. Feeling guilty for this, hating to fill his mind with information which he had not chosen for himself, he crammed intensively for two months before the final examination. So painful did he find this enforced study, so keen was his reaction to it that he said, ". . . after I had passed the final examination, I found the consideration of any scientific problems distasteful to me for an entire year."

At about the time he began working at the Swiss patent office his interest in physics had revived and there, as we have seen, he had formulated the statistical mechanics which made it possible to account for Brownian motion and he had worked out the photoelectric theory. There, also, he had completed his first theory of relativity which, because it considers a special case—motion in a straight line at unvarying speed (uniform motion)—is called the "special theory of relativity." Building on this, Einstein later found a way to account for nonuniform motion as well, in his general theory of relativity. These theories are known best for certain consequences which Einstein was able to deduce from them. From the general theory came deductions as to the size and structure of the universe; from the special theory came the equation $E = mc^2$, which played a decisive role in the investigation and development of nuclear energy.

The line of reasoning which led to that theory began when Einstein was sixteen years old. He was "wondering," wondering about motion and light. The speed of light had been clocked; in a vacuum it travels approximately 186,000 miles every second. (This means that light can travel around the earth in a seventh of a second.)

What would happen, Einstein wondered, if an observer could travel as fast as he liked, even as fast as light? Suppose I were running almost that fast and I chased a light beam? Its speed in relation to me would be much less than 186,000 miles per second. And at the same instant that light moved slowly for me, another person who was standing still and measuring the speed of the beam would clock it at 186,000 miles per second.

This conclusion seemed wrong to him somehow, but there was nothing in physics which denied it. He carried the reasoning a step further: suppose, he thought, that I were running at the *same* speed as light. Suppose I were riding on the beam. Then, for me, light would be at rest. A beam of light at rest. How could there be such a thing? Light was defined in terms of its frequency of *motion;* light at rest was a contradiction in its own terms—a paradox.

These thoughts, this wondering which he experienced at sixteen, made him feel uneasy, Einstein said later—but also excited. Then, when he was a student at the Zurich Institute, he read about some experiments which had been done in an attempt to solve a closely related problem. They had been designed to measure the speed of light in relation to a substance which was supposed to permeate all of matter and space, the ether. This substance never had been detected in an experiment. It *was,* however, a fact of experiment that light is transmitted in the form of waves and that these waves could travel through space empty of air molecules or any other known form of matter. All other wave phenomena required a medium of some kind to act as carrier; light also, it was assumed, required such a carrier. This was the function of the ether, which was assumed to be so fine a substance that it escaped detection.

If the ether-substance which filled the universe was stationary, then the planet Earth, moving through it, would meet resistance and a current, a "wind," in the ether would be created. A light beam sent against this current ought to be slowed down to a certain extent. Light sent in the opposite direction, with the current, ought to move just so much faster. An American, A. A. Michelson, decided to test this assumption in an experiment. He invented an instrument, fundamentally a system of mirrors which split a light beam, sending it at the same instant in different directions, and reflected the beams back to the source for observation. Michelson's first test was made in Germany in 1881, when he was studying in the laboratory of Hermann von Helmholtz. The results were negative: the travel time of the two light beams was not different. Both were clocked at 186,000 miles per second. Six years later, in the United States, Michelson repeated the experiment, this time collaborating with E. W. Morley and using a more refined instrument. The result was the same.

This did not necessarily mean that there was no ether. It was possible that the ether was not stationary but was dragged along with the earth, in which case a current would not be created and the travel time of the two beams would not be affected. This was one hypothesis; there were others.

Einstein, when he heard of the experiment, interpreted it as bearing on the paradox which had occurred to him years before and which he had not ceased to ponder. Putting aside the question of whether there was or was not an ether, the experiment demonstrated that the speed of light always was the same relative to the earth, although the earth continually changes direction as it orbits the sun. One could conclude from this that no matter how fast an observer might travel, he never would catch up with light. He never would observe it at rest.

Now Einstein had support for doubts which he had felt much earlier. Now he was confronted with understanding the constant speed of light. *How could it be that at the same instant of time light travels at 186,000 miles per second both for an observer who is moving in the same direction as light and for an observer who is standing still?*

When Einstein at last found a way to answer this question, he had the special theory of relativity, at least its essential features. In order to answer it, he did not need any new information; it was a matter of spotting a flaw in the question, something it took for granted for which there was no justification. (The reader, who has the advantage of living after Einstein, may be able to recognize this flaw.)

Not knowing that the clue lay in the question itself Einstein, at the Zurich Institute and afterward, explored the known facts of physics, seeking a point of departure, something on which to build. He could not find it. Passionately, he wanted to understand. It was as if he were being driven toward something, he said, but could not reach it. He was depressed often; at times he felt despair.

As he studied physics independently (and without regard for the examinations) he became more critical of it. It was not simple enough; it was built upon assumptions which ought not to be necessary—the ether, for instance. It had been invented so that light might be understood in terms of a supporting medium. But there was no experimental evidence for the ether. Perhaps it was no more than an invention which helped one to understand.

There was a resemblance between this hypothetical ether and another idea, an idea which had been introduced into physics centuries earlier in order to account for motion.

In the universe there is no planet or star which can be used as a fixed reference point for judging the motion of something else. All is in motion. The earth rotates on its axis as it moves in an orbit about the sun. The sun, together with our solar system, is moving within the Milky Way galaxy, and that galaxy itself moves, relative to other galaxies. Long ago the question arose, "How can one determine the true, the absolute, motion of a body when it moves differently in relation to the earth than it moves in relation to the sun?" This question could be answered if space was conceived as a fixed container for the celestial bodies, a container which was not influenced by their motions. Thus space became a reference frame, making it possible to assign absolute motions to the celestial bodies as to any other object.

Einstein questioned this idea of fixed—or absolute—space just as he questioned the ether idea. There was no experimental evidence for either one and as a physicist he believed in the rule which says, "If you cannot relate it to a measurable quantity, you do not know it." He felt, moreover, that these mental constructions should not be necessary in order to understand. Fundamentally, the pattern must be a simple one. "God is sophisticated but He is not malicious." Thus it ought to be possible to deduce Newton's laws governing motion *without* using Newton's contrived assumption of absolute space.

In his thinking, Einstein was greatly influenced by the philosophical writings of Ernst Mach, who criticized the ideas on which classical physics is built, criticized them as insufficiently related to experimental findings. Not only did Mach regard the idea of absolute space as questionable, he also criticized the concept of absolute time, time as we think of it in everyday life.

It was just this idea which had blocked Einstein in his attempt to solve the problem posed by a constant light speed. The turning point in his long search for a solution came when in line with the ideas of Mach he began to wonder whether the everyday conception of time could pass the measurement test.

What is a clock? he asked himself. What do we do when we "keep time"? Can we do this under all circumstances? Do we know that our timepieces *always* keep the same rhythm? Suppose a clock were moving at an enormous speed, a speed comparable to that of light. Do we know that the clock's rhythm would not be affected then?

After making an analysis of the experiments which could be done to answer this last question, he concluded that the answer had to be, "We do not know." It was just as reasonable to assume that a clock's rhythm was affected by motion as to assume that it was not.

Suppose, then, that different observers moving at different speeds did not reckon time the same. Was there any possible means of comparing the different reckonings? Could the time difference be allowed for? Analysis of possible experiments gave the answer: No. It could not be.

Now we can see how Einstein solved the problem which we stated earlier: "How can it be that at the same instant of time light travels at 186,000 miles per second both for an observer who is moving in the same direction as light and for an observer who is standing still?"

The question assumed that different observers moving at different speeds were able to make the same reckoning of time so that they measured the speed of light at the same instant. There were no firm grounds for this assumption. Could it not be, then, that estimates of time by different observers varied with their speed of motion, varied in just such a way that the speed of light always must be *observed* as constant?

This is the primary idea of Einstein's special theory of relativity. It had taken him almost seven years to get to it. The notion of absolute time was "anchored in the unconscious," he said, and very hard to question. But once he did, the rest came quickly. It took him only five weeks to translate the private sign language of his thoughts into precise terminology and trace out logical consequences.

The result was a simple theory, in the sense that much followed from little; and this little, these few principles (axioms) with which he began, had a firm base in experiment. The cardinal axiom of the theory was simply that the speed of light always is observed to be the same, regardless of the (uniform) motion of its source or of its receiver. He began the theory by saying in effect: "Let us see if we can do without such ideas as ether, absolute space, absolute time. Let us take nothing for granted but this one axiom and see what we are able to deduce from it."

In Einstein's deductions the speed of light (which physicists represent with the letter "c") appears often, since he was using it as an organizing principle. It appears in the new laws governing moving bodies which are a part of the theory, laws which agree with those of Newton for bodies moving slowly relative to c, but differ with Newton's when their speed approaches closely to c. According to these laws mass increases with speed, becoming infinitely great at c. Therefore nothing in the universe can travel as fast as 186,000 miles per second.

It follows also from these laws that mass and energy, hitherto considered separate, are in fact different aspects of the same thing. A tiny mass is equivalent to a vast amount of energy, or $m = \dfrac{E}{c^2}$ which also may be written $E = mc^2$. This equation did not, as sometimes is claimed, reveal the atom's huge store of energy. Rutherford and others who studied radioactive decay were well aware of it. The equation did form a quantitative basis for that energy's later exploitation.

For Einstein $E = mc^2$ had a special meaning. The equation stated a fundamental relationship in nature and the equation followed by logical deduction from assumptions which, as Einstein saw it, were simpler than those of Newtonian physics. Here then was the justification for his belief in an underlying universal pattern which man could discover because of his notion of logical simplicity.

A few years after the special theory of relativity was introduced, a mathematician, Hermann Minkowski, recognized that the theory could be expressed in a different mathematical form from the one Einstein had used. In Minkowski's mathematical translation of the theory, space and time appeared as a unity, a four-dimensional continuum. Einstein had shown that an event must be measured differently by different observers. But the theory also had provided a means of correlating divergent observations so that one could arrive at a reliable measurement, a measurement valid for all observers. This was done, Minkowski showed, by referring data obtained from observation to a mathematical space-time framework. Thus it was revealed that Einstein's theory introduced a new definition of space and time.

The problem now was to fit all motion into this four-dimensional scheme, not only the special case of uniform motion, and thus arrive at a general theory. Einstein began to work on this.

Toward the end of his life, in a chat with Robert Oppenheimer, Einstein spoke of how his great work, done when he was twenty-six, had affected the rest of his life. "When," he said, "it has once been given you to do something rather reasonable, forever afterward your work and life are a little strange."

In the years following 1905, Einstein would become a symbol to different people of different things. To some he would appear a dangerous radical; to others, a naive incompetent. Some understood his work to mean that every absolute, including "good" and "evil," had been abolished. Others would see him as a religious leader who in the manner of Moses helped lead his people to a new land.

In 1905, Einstein's fame and its strange consequences lay far in the future, but almost immediately after his work became known to other scientists his life began to change. The bars to the academic world which he had encountered a few years earlier when seeking a position were dropped; he was urged to enter. Soon after his work of 1905 appeared, Einstein left the government office where he had been obscure and happy.

Seven years after he became *Privatdozent*, at the University of Berne, he had reached the summit of his profession and was, at thirty-three, a full professor. The customary slow progress from rung to rung of the academic ladder had been accelerated, in recognition of his ability. He had been "called" from the University of Berne to that of Zurich; then to the University of Prague and back to Zurich again, to the Polytechnic Institute where once he had been a student. Each move meant a higher rank in the academic world (usually a larger salary) and increased respect.

Each elevation made Einstein feel a bit more uncomfortable. At the patent office he had taken a certain satisfaction in completing the duties expected of him. At the university, this was not the case. He avoided his academic chores or performed them sketchily. As a consequence, he felt that he was not worth the salary his university-employer paid him, or rather he was worth it only if his research led to results which would serve to enhance the university's reputation. He was, he felt, being paid to have ideas. As in his student days, others had a claim on his thoughts. He lived with an uncomfortable sense of obligation and thought wistfully of the happy lives of shoemakers and lighthouse keepers who were paid for performing simple tasks, tasks which left their minds free.

In the year 1913 Einstein, in Zurich, received a visitor from Germany who had come on an important mission. It was Max Planck, one of the first physicists to recognize the importance of the special theory of relativity. "If [it] . . . should prove to be correct, as I expect it will, he will be considered the Copernicus of the twentieth century," Planck had predicted in 1910. He admired the theory of relativity not because it challenged ideas which hitherto had been regarded as absolute but because the theory, as he saw it, introduced a new absolute into physics. "Everything that is relative," said Planck, "presupposes the existence of something that is absolute, and is meaningful only when juxtaposed to something absolute." In the case of relativity, the absolute was the four-dimensional space-time continuum.

Planck had come to Switzerland to offer Einstein what was probably the best position in all Europe for the kind of physicist Einstein was. At a new, richly endowed research center he would direct the work of outstanding physicists. At a great university he would have the title of "professor," but no academic duties to perform. Unless he wished to he would never have to lecture. His salary would be quite large; his freedom quite unrestricted.

It was a very attractive invitation, but it was an invitation to Germany: to the research institute founded by and named after the emperor, Kaiser Wilhelm II; to the university of von Helmholtz and Planck, the University of Berlin. Einstein, needless to say, did not like the idea of returning to the country which as a youth he had been so happy to leave, as well as to the very center of that academic world which he found so oppressive. Even with Planck, whose dedication to physics so resembled his own, he felt a little ill at ease. There was something about the serious, formal Prussian manner which always put him off.

On the other hand, if he accepted Planck's invitation he would mingle with some of the greatest scientists of his time, which would stimulate his own thinking. Unencumbered by lecturing he would be free to concentrate on his own work, the extension of his first relativity theory. Above all else, he wanted the opportunity to work on this.

The two physicists sat together and talked about Einstein's future—one a gaunt, serious man with a restrained, formal manner; the other a bit plump, with sad, bright eyes and unruly hair, a man who made jokes and laughed often.

On that occasion Einstein did not reply to Planck's offer. Ultimately, however, he found it irresistible and just before the beginning of the First World War he returned to Germany. It was a decision which would contribute to what he called the strangeness of his life.

Einstein returned to Germany in the year 1913. During the same year Niels Bohr, the young Dane who had been studying with Rutherford in England, found a way to answer the questions posed by an atomic nucleus. The solution he found stemmed from ideas of Planck and Einstein.

In the chapters to come, we will follow the work of Niels Bohr and of certain other young physicists who, together with Bohr, worked on atomic theory. As in the previous chapters, we will be concerned with how they worked as well as with what they achieved, with their different ways of doing physics, their differences as scientists and as men.

After we have seen Bohr's atomic theory develop and change to become the one of present-day physics, we will return to Albert Einstein, to his life after he went back to Germany and then to his debate with Bohr about the meaning of the new atomic theory.

Now, in 1913, the road of Planck and Einstein joins the road of Rutherford to become an arterial highway, as Niels Bohr discovers how to explain the impossibilities of Rutherford's atomic model.

CHAPTER SIX

Niels Bohr: Early Quantum Theory
of the Atom

> *. . . this is the day we celebrate Bohr*
> *Who gave us the complementarity law*
> *That gives correspondence (as Bohr said before)*
> *That holds in the shell as well as the core*
> *That possesses the compound levels galore*
> *That make up the spectrum*
> *That's due to the modes*
> *That belong to the drop*
> *That looks like the nucleus*
> *That sits in the atom*
> *That Bohr built.*
> —by R. E. Peierls, to celebrate
> Niels Bohr's seventieth birthday

THIS STORY is about a new and good Elephant's Child, who was never spanked and who never lived in Africa. . . . This Elephant's Child had a very remarkable nose right from the beginning . . . and he asked ever so many questions . . . new and unheard of questions."

So begins a fable told by a physicist, Oskar Klein, about Niels Bohr, a fable inspired by one of Kipling's *Just So Stories* about the Elephant's Child who *was* spanked by his family for asking questions and learned the answer to one from the Crocodile in the great greasy Limpopo River, in the process of having his nose pulled into a trunk. The new just so story goes on to relate how Bohr, the Elephant's Child, went to "the great greasy town of Manchester" to find Old Crocodile and ask him questions about the atom and how the Elephant's Child, having found the answers to his questions for himself, grew up to become a leader and was loved and honored by all in his land.

Niels Bohr would become the first citizen of Denmark. In that country he would be accorded the kind of respect which in the United States sometimes is given to great military leaders. Often he would be called upon not only to direct scientific programs for Denmark (he was head of the Danish Cancer Committee as well as the Danish atomic energy program) but to perform such civic chores as judging a competition or raising funds for a museum of art. Bohr was happy to do such things. Unlike Einstein he did not feel isolated from the society into which he had been born. He was not an outsider.

Another striking contrast between these two men was in their physical appearance. Einstein, particularly when he was older, looked like a person given to deep, profound thought; his face perfectly expressed what he was. Bohr's face, on the other hand, had a heavy, sagging look; his small eyes were set closely together, his cheeks drooped like a hound's, his lips were large and thick.

Bohr had a brother Harald, who looked very much like him and later became an outstanding mathematician. Once when the two were young, they were riding with their mother in a Copenhagen streetcar. To pass the time, she told them a story. Another passenger on the streetcar, watching the boys as they drank in the story, their faces blank, their jaws probably sagging, was heard to remark, "*Stakkels Mor,*" Danish for "That poor mother."

Niels Bohr did not look intelligent. Also, unlike Einstein who possessed a flair for using words and expressed his thoughts easily, clearly and vividly, Bohr spoke tentatively and it was

The sons of Professor Christian Bohr. Harald Bohr on the left; Niels Bohr on the right.

hard at times to make out his meaning. This was only partly due to the fact that his voice was soft and that he had a slight speech impediment. There was also the fact that he did not necessarily try to express his thoughts in the clearest possible way. For Bohr words were tools: in doing physics he used words almost as much as he used mathematical symbols. Often when he talked he was not reporting a conclusion but working toward it as he spoke. Once one got to know Bohr and understand his way of using words, conversation with him could be exciting, especially if one questioned his ideas. In argument he was at his best.

But at the Cavendish Laboratory, where he had gone to study under J. J. Thomson, no one spoke his native Danish and Bohr was having trouble with the English language, so that he was even more unintelligible than usual. His associates were mystified by such things as the Dane's frequent use of the word "load" in connection with the electron; and at times he appeared to be speaking of a person, someone called "Zha-a-a-n." Some time later Bohr discovered that the English word for "charge" was *not* "load" and that in England one did not give the French-looking name of the physicist James Jeans the French pronunciation.

When he recognized that he was not making himself clear to the Englishmen, who spoke so precisely themselves and were extremely conscious of grammar and accent, Bohr bought a dictionary and the entire collected works of Charles Dickens. Systematically he began to read the set of volumes, starting with the first, and as he read these English classics he looked up every word of which he was unsure in the English dictionary.

It may have been due to faulty communication that J. J. Thomson, who had been quick to recognize Rutherford's ability, failed to notice Bohr's. Once, at a meeting where the young man had modestly advanced an idea to the assembled scientists, Thomson had broken in to remark hastily that Bohr's idea was useless—and then had gone on to say the same thing in different words.

Many years later, Bohr described his first impression of Ernest Rutherford at the ceremonial and noisy Cavendish dinner, where he first saw the older man. Bohr said that he especially had liked the enthusiasm with which the New Zealander spoke of another's work (it happened to be the cloud chamber of C. T. R. Wilson). Not long afterward, the young Dane had asked if he might join the Manchester group—"to get some experience," as Rutherford put it.

For Bohr had come to Manchester expecting to experiment.

Like Rutherford he enjoyed using his hands and was good at it. In Denmark he had won a gold medal for some of his experiments. Yet, as earlier noted, it was not the experimental possibilities opened up by the discovery of a nucleus which had caught Bohr's interest but the theoretical problems it presented. Why did the electron, the unit of electricity, fail when inside the atom to obey the laws of electricity? Given a nucleus of opposite charge, the electron should be attracted to it and move in an ellipse, like a planet circling the sun. Moving in this way, the electron should constantly emit electromagnetic radiation, lose energy and rapidly spiral into the atom's core.

As in the case of black-body radiation, physics prescribed catastrophe. For the ultraviolet catastrophe Max Planck had found a "cure." By restricting the energies which physics said were possible, by means of his quantum equation $E = hf$, he had been able to bring theory into line with fact. Perhaps a further application of the quantum idea would solve the atomic problem also. Perhaps there were other restrictions on energy, restrictions which explained why matter under ordinary conditions does not glow with light and why the electron does not fall into the nucleus.

Niels Bohr thought so. He was one of the small group of physicists who recognized Planck and Einstein's quantum idea as evidence of something new and far-reaching rather than a clever "German invention" which would not survive. At Manchester Bohr looked for a way to account for atomic behavior by a new application of the quantum idea. He could not do it. What he lacked (although he did not know it) was certain information. This data already existed—and had for several decades—but Bohr was not very familiar with the specialized science called "spectrum analysis" or "spectroscopy." He did not know about the formulas which had been worked out in connection with what are called "line spectra."

Spectra of this type are observed when matter in gaseous form is heated to a high temperature until it glows with light. With a spectroscope the light then is separated into its component colors, which is to say wavelengths or frequencies. While the spectrum of a heated glowing solid (such as the black body) contains virtually every frequency from red to violet and resembles a ribbon of continuous color, one hue blurring into another, the spectrum of a heated gas shows only a limited number of frequencies. These appear in the spectrum as separate lines of color—hence the term "line spectrum." Each chemical element in gaseous form has a different spectrum. In the visible part of the hydrogen spectrum, for example, there are just three lines: one red, one green, one violet; and only hydrogen has this particular spectrum.

In 1912, when Niels Bohr was trying to understand the solar-system model of the atom, the science of spectroscopy was much more isolated from the rest of physics than it is today. While it was thought that the line spectra must reflect atomic behavior in some way, attempts to demonstrate this had ended in failure. For several decades the spectroscopists had been photographing or diagramming the line spectra and finding ways to describe their regularities with mathematics—the sum of this information being a stack of heavy volumes. Many physicists considered the job of relating these often very complicated spectra to the atom insuperable; at least one physicist, Niels Bohr, was totally ignorant of the formulas of the spectroscopists. Years later he said that in the old days the line spectra had seemed like the wing of a butterfly—complicated, colorful, pretty and just about as significant as the wing of a butterfly. Bohr, to repeat, was lacking in the knowledge necessary for a quantum theory of the atom.

At the end of the spring term, in 1912, the fellowship money which had helped to pay Bohr's way in England had been used up and, his mind busy with the unsolved atomic problem, he returned home to Denmark, where he began to teach at the University of Copenhagen. Though his stay at Manchester had been brief, less than half a year, he felt that he had gained much from it and in a letter to Rutherford he struggled with his limited English to tell the older man what he felt:

> Leaving Manchester I wish to thank you very much for all your kindness against me here. I am so glad for all what I have learned during my stay in your laboratory, I am only sorry that it has been so short. I am so thankful for all the time you kindly have given me; your suggestions and criticisms have made so many questions so real for me, and I am looking forward so very much to work upon them in the following years.

The peninsula called Denmark to which Bohr returned borders on Germany and superficially the two countries then appeared much the same, with their small, tidy farms worked by solid-looking men and women, and their medieval towns with narrow streets of cobblestone. But the emphasis on rank, the love of order and reverence for authority, so much a part of German life, were far less pronounced in Denmark.

In Copenhagen, his birthplace, Niels Bohr had grown up in an atmosphere of discussion and speculation. His father, Christian Bohr, was a professor at the University of Copenhagen who taught physiology and was greatly interested in philosophical questions connected with the scientific study of life. He allowed his sons Niels and Harald to use his laboratory and taught them experimental techniques. He was interested in their ideas; indeed he called his son Niels "the thinker of the family."

Every other Friday, Professor Bohr's friends—a philosopher, a physicist, a scholar of languages—would come to his home for dinner. Afterward the four professors liked to explore general questions together, each contributing from his own specialty. They speculated about the unknown. The sons of Professor Bohr listened.

Later Niels Bohr took a course in philosophy at the University of Copenhagen, a course taught by his father's friend Harald Høffding, who was a distinguished philosopher. Bohr was troubled by some errors in the lectures. He spoke to the professor about mistakes he had made in logic, whereupon the great Professor Høffding, after correcting his work, submitted it to his young student for approval.

From boyhood Niels Bohr had been encouraged to follow his natural bent for speculative thought. But it would be wrong to think that his youth was spent in passive cogitation. He was strong, energetic, good at sports. He played football passionately and became a member of Denmark's champion amateur team. He was a proficient and powerful skier. Friends remember him as always moving fast. As a young professor of physics, he bicycled to his office, like many other Danes, and "would come into the yard, pushing his bicycle faster than anybody else." A serious discussion of physics with a colleague would be "interrupted by short running trips."

And while Bohr was a serious thinker, he was not given to solemnity. "There are," he would say, "things so serious that you can only joke about them." At a time when the foundations of physics were shifting, when problem after problem could not be solved and no one knew why, when physicists talked late into the night testing solutions, finding none, Niels Bohr might invite young colleagues to the Tivoli, an old amusement park at the center of Copenhagen where as well as taking the usual rides one could watch puppet shows and operas, magicians, parades and fireworks. Once the group of young physicists stopped to watch a demonstration of so-called "thought transmission" and Bohr became very interested. Soon he had thought up a theory, based on ventriloquism, to explain what was happening in the show. He expounded this theory at length with all the "scientific" trimmings. He met the objections of his colleagues with skill, heat, eloquence. It was the same sort of discussion which kept them all up late into the night. But now Bohr was playing.

This serious yet lighthearted young man, this athlete who loved poetry and painting, sculpture and music, who possessed the strange aspect of an Elephant's Child but who was not spanked for asking questions, had begun as a schoolboy to ponder ideas which would play a part in his later work—in his atomic theory.

While adolescents are apt to speculate about "What is the meaning of life?" "Why are we here?" Niels Bohr was preoccupied with a different sort of question: "How can such words as 'existence' and 'reality' be used meaningfully when they represent different things to different people at different times?" It seemed to him that the actual experiences which lay behind such word-generalizations were so many, various, even contradictory to one another, that logical debate centering on these words was useless and empty. There was, for instance, the philosophical debate as to whether man's choices are dictated by his biological nature or made freely. Such a question stated two opposite generalizations: free will and determinism. Bohr thought that these two abstractions were not irreconcilable opposites. They simply expressed different aspects of man's actual situation in life. "A common human experience," he would say, "is the feeling of . . . being able to make the best out of circumstances." In this feeling which unites the ideas that choice is free and that it is restricted Bohr saw apparent opposites resolved and in harmony.

Such speculations seem far afield from the science of physics, yet they did play a part in the way Bohr attacked scientific problems. For example, the failure of physics when applied to the atom could be interpreted in different ways. It might mean that established or classical physics was entirely wrong. Or one could interpret this failure to mean that essential information was missing: future atomic experiments would turn up new facts revealing that classical physics was not fundamentally wrong but only had been misapplied.

Bohr did not draw either of these conclusions. He thought of Einstein's special theory of relativity, of the motion laws which replaced those of Newton for bodies moving close to the speed of light but which gave the same solutions as Newton's laws for the lower speeds. Again he knew that while Planck's quantum rule for black-body radiation disagreed with classical law in the region of short waves and high frequencies, for the long wavelengths and low frequencies the contradictory laws were in close agreement.

On the basis of this evidence Bohr thought it likely that there would someday be a new physics broad enough to contain the old within its scope. Classical physics would continue to hold where it had in the past; the future would reveal it as part of a larger theory. In other words Bohr believed that what appeared to be irreconcilable opposites might well be different aspects of the same thing and, seen in a wider context, prove harmonious.

As we will see, it was just this idea which would help him to work toward a new atomic physics. He would seek a connection between apparent opposites: after making a hypothesis to explain the atom's mystifying behavior, he would extend this hypothesis beyond the limits of the atom to learn whether or not it approached agreement with a classical law, where that was known to hold. Working in reverse, he would apply a classical law to the case of the atom. Even though this gave a wrong solution, it was not necessarily wrong in every detail. At times the wrong answer could provide a crude hint, a clue.

Yet neither this tool, which Bohr would call "correspondence," nor the quantum idea could serve him until he learned about the spectroscopic data mentioned earlier. Back in Copenhagen, after leaving the Manchester group, he had continued with his attempt to solve the problem of atomic structure, finding it impossible, as before, to start an experimental investigation. Instead, using his own term, he "dreamed" about the problem. It was always on his mind.

The day in the early spring of 1913 when Niels Bohr finally happened upon a formula which described the line spectrum of hydrogen was a turning point in the history of physics comparable to Isaac Newton's discovery of his gravitation law, and as in the case of Newton there is a fable to mark it. Newton's idea, it is rumored, came to him when he was struck by an apple which fell from a tree; Bohr, so the rumor goes, was similarly "struck" when he chanced to pick up a book in which the crucial formula appeared, a book not for grownups, but for children—a juvenile book.*

The actual facts are less colorful: One day Bohr was talking with a colleague about the unsolved problem, a colleague who specialized in spectroscopy. From him Bohr got the idea of looking up some of those heavy volumes of spectroscopic data, and there he found the hydrogen formula. It had been worked out by a Swiss schoolteacher, Johann Jakob Balmer, in 1885, the year of Bohr's birth. Twenty-seven years later, when Bohr saw it, everything began to fall into place.

Balmer had been one of the first scientists to become interested in the numerical relationships of spectral lines, with their butterfly-wing regularities. He had thought about the three lines in hydrogen's spectrum, pondering the known fact that the wavelengths of the red and green lines were related to each other exactly as the whole numbers 27 and 20 are related, while the

---

* The Balmer formula was part of a physicist's education then as now and it is hard to believe that Bohr never had encountered it. Perhaps the myth of the children's book has spread among physicists because of their amazement that he could have missed what was such common knowledge that one *almost* could say children knew about it before Bohr did.

green and violet lines were as 28 to 25. At length he had arrived at a formula which amounts to a number trick, giving, in mathematical shorthand, these instructions:

> Square the number 3. Divide 1 by the result and subtract this fraction from ¼. Multiply the answer by the number 32,903,640,-000,000,000. This gives the frequency of the red line in hydrogen's spectrum (or the wavelength for one is easily derived from the other). If you begin with 4, you will obtain the frequency of the green line and 5 gives the violet line's frequency.

Only three lines had been detected in the hydrogen spectrum when Balmer worked out this formula. Later other lines were found and their frequencies could be calculated from the same formula, using the numbers 6, 7, 8 and so on.

Like Planck's first formula for the energies of black-body radiation, Balmer's was an empirical formula: it summed up observations but it did not account for them. And the Swedish spectroscopist J. R. Rydberg, who used the same number 32,903,-640,000,000,000 in formulas describing the spectra of other elements, did not know why his number tricks worked either. Because of his accomplishments, the number was called "Rydberg's constant," and as will be seen it played an important part in Bohr's work.

But first, what did Bohr see in the Balmer formula? In what way was it a clue to atomic structure? The answer is quite simple: Bohr recognized that the Balmer formula could be written in a slightly different way, using the symbol h, Planck's constant. This gave him a formula which described the actual spectrum of hydrogen, as determined by experiment, just as did Balmer's, but a formula which also *accounted for this spectrum on the basis of energy quanta.*

As we have seen, physics when applied to the case of the atom gave impossible conclusions. Now, starting with the Balmer formula, Bohr put together a quantum theory of the atom which accounted for the fact that atoms do not ordinarily emit radiation and that the electron does not spiral into the nucleus. Bohr's theory did not lead to impossible conclusions. Not unexpectedly, it came into conflict with the ideas of classical physics. Below we list the conflicting ideas:

CLASSICAL: The electron circles the nucleus in orbits of varying shape and size. In each orbit its energy is a little different. The possible orbits, which are the same as the possible energies, are infinite in number: energy is continuous.

BOHR: The electron does not move in all the orbits which are possible but only in those where its energy is proportional to Planck's constant multiplied by one of the whole numbers. Energy is discontinuous.

CLASSICAL: The electron, attracted by the nucleus, revolves about it radiating energy and spirals into the atom's core.

BOHR: The electron may shift from one of the specified quantum orbits to another but these orbits represent its *only* energies. The electron may go no further toward the nucleus than the innermost quantum orbit.

CLASSICAL: The electron moving within the atom always emits radiation and the frequency of this radiation is the same as the "oftenness" with which the orbit is completed.

BOHR: The electron moving in a quantum orbit does not emit radiation. When enough energy is added to the atom (by heating, for example) the electron is forced outward from an inner orbit, associated with low energy, to an outer high-energy orbit. Radiation is emitted as the electron, attracted by the nucleus, falls back to an inner orbit. The jump from high energy to low produces a definite portion of light energy. The frequency of this is equal to the difference in energy between the two orbits, di-

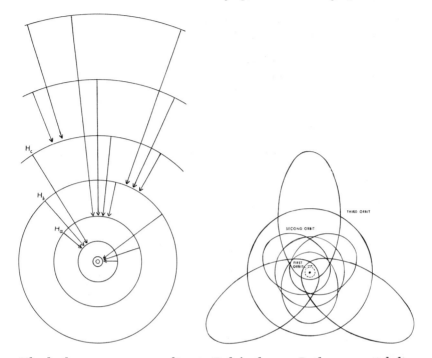

The hydrogen atom according to Bohr's theory. Both are partial dia-grams. The larger one illustrates some of the jumps between quantum orbits which give rise to the different frequencies in hydrogen's spec-trum. The three lines in its visible region, known as "the Balmer series," are labeled H_a, H_b and H_c. The smaller diagram is a slightly different (and more familiar) representation of the Bohr hydrogen atom.

vided by Planck's constant. This is simply another way of stating Planck's equation $E = hf$. The "E" becomes "loss of energy" and the equation, turned around, then states:

$$\text{frequency} = \frac{\text{loss of energy}}{\text{Planck's constant}}$$

The quantum jump of the electron from orbit to orbit means that energy is sent out (and absorbed) in quanta. Applying the idea of Max Planck to the case of atomic structure, Niels Bohr had found a way to account for known facts of atomic behavior—but this did not mean that his theory necessarily was correct. There might be another way to account for these facts. There was no experimental evidence that the electron actually moved as Bohr had stipulated; there was no observational support for the idea of quantum orbits. They had been "invented" to account for what already was known. It was true that on the basis of his assumptions Bohr could account for the actual hydrogen spectrum (see diagram), but even this success was not as great as it might at first seem. The theory had been designed in the first place to account for the hydrogen spectrum when Bohr wrote Balmer's formula in terms of Planck's constant.

In the next chapter we will describe the experiment which tested some of Bohr's ideas, but even before this experiment was performed there was other evidence that he was on the right track. As we have said, the theory yielded facts for which it had been designed, but it yielded something else as well—something which it had not been designed to do: Bohr's theory revealed the meaning of that mysterious large number, Rydberg's constant.

In finding that meaning Bohr had employed the method he later called "correspondence." According to that idea an hypothesis concerning the atom should in its broadest extension move toward agreement with classical physics. Applied to his quantum hypothesis of the atom, this extension would be a quantum orbit very far from the nucleus and therefore enormous in diameter, as large as a radio antenna. (An orbit of this size could occur in nature only if atoms were widely separated from one another. On earth the proper conditions for such a separation do not exist; that they do on some stars is indicated by spectrum analysis of starlight.) For such a large orbit Bohr found that the predictions of his theory did indeed move toward agreement with the classical laws of radio and other large-scale electromagnetic phenomena. With this established, he used the classical laws to develop his theory further and thus was guided to a formula giving the energies of the various quantum orbits, and there in the formula appeared the following symbols:

$$\frac{2\,\pi^2\,me^4}{h^3}$$

Here m stands for the electron's mass, e for its charge and h for Planck's constant. The numerical values of these three are known from experiment and when they are substituted for the letters and the formula is solved, one arrives at the following number: 32,903,640,000,000,000 vibrations per second. This is Rydberg's constant, the fixed number which made the arithmetical tricks of Rydberg and Balmer possible. On the basis of his quantum assumptions and by using his correspondence idea Bohr had found the ingredients of that number and therefore why the number tricks worked. This was very impressive evidence for his theory.

When Niels Bohr completed the paper in which he presented his atomic theory he sent it to Ernest Rutherford in Manchester. Before the new ideas could appear in print, Rutherford's approval was necessary.

Once before, during his stay at Manchester, Bohr had gone to Rutherford with a theory. The young Dane had seen a connection between the discovery of a nucleus and the laws of radioactive change worked out prior to that discovery. Could it not be that the emission of alpha and beta particles meant a change in nuclear charge? That the products of radioactive change were new elements because of this change in the nucleus?

As head of the Manchester laboratory Rutherford had to approve these ideas before they could be submitted for publication and he had advised Bohr to wait until there was more experimental evidence to go on. Then a few months later similar ideas had been published by other physicists (Soddy and Fajans). When verification came the credit was theirs.

With this as background a sentence in the letter Bohr sent to Rutherford with his paper takes on added significance: "I am," the young man wrote, "very anxious to know what you may think of it all."

Bohr had worked hard on his presentation of the new theory. Writing down his thoughts, putting them into final form, was difficult for him. In the first place, he needed a listener, someone to tell his ideas to. When he was a boy his mother had served this function, helping him with homework, and later other physicists would act as his sounding board and take down his words. (Rarely did Bohr touch pen to paper; his handwriting was totally illegible.)

But even with the necessary help, he resisted completing an article or scientific paper. Many physicists feel that their work is done when they have shown the basis for their ideas logically and clearly; Bohr did not share this feeling. An enlargement of knowledge meant above all that there were more questions to ask. When he wrote he wanted to show not just how things did fit together, but also how they did not fit, which pointed the way to further progress. He did not want to be *too* clear, for that meant leaving things out, often just the things that led to a deeper understanding.

In the paper he sent to Rutherford he had wanted to underline the contradictions between his quantum ideas and the ideas of classical physics, because he thought that progress could be made by stressing these and working with them in the sense of correspondence. This idea, which on the face of it sounded illogical, was difficult to communicate. Later when some of Bohr's writings were translated into other languages, the words "Just by accentuating this contrast, it will perhaps be possible . . . to bring about a certain consistency" were changed by two different translators, presumably because they thought that this could not possibly be what the author had meant to say.

Across the North Sea in Manchester Rutherford received Bohr's paper, his first major work in physics, the result of many drafts, revisions, sleepless nights. The paper contained numerous references to the work of other men, in the attempt to give credit where it was due—indeed, far more references than were necessary, Bohr said later. And in addition to the other difficulties it presented to the reader, the wording was awkward: the young Dane still was having trouble with English. ("It would be better," Rutherford told him later, "not to start every sentence with 'however.'")

Rutherford liked things to be simple. He said once that if a theory were any good it ought to be possible to explain it to a barmaid. When he replied to Bohr about the paper, he said that the ideas were "very ingenious and seemed to work out well." But then he went on: "I think in your endeavor to be clear, you have a tendency to repeat your statements in different parts of the paper. I think that your paper really ought to be cut down and I think this could be done without sacrificing anything to clearness."

This, said Bohr later, "brought me into a quite embarrassing situation." Not satisfied with the paper he had sent to Rutherford, he had rewritten it. The new version was much longer— and he had already mailed it off to Rutherford without waiting for his reaction to the first version.

Now Bohr decided that the best way to straighten things out was to go to Manchester and talk the whole thing over with Rutherford in person. He got on a boat to England.

On the day he arrived at the Rutherfords' home, they were entertaining a visitor in the parlor, a visitor who later remembered the "slight-looking boy" who arrived suddenly and was taken at once by Rutherford to his study.

There the two remained for a long time and the next evening

found them in the same place arguing about the same things. Although Bohr was a modest person, he was not shy about defending his ideas. Indeed, he thrived on opposition. No longer did he speak tentatively but with spirit, vigor, eloquence. A friend has said that Bohr even looked different when he argued: the heavy, irregular features seemed to fade away; one noticed above all the eyes "under the shield of the thick eyebrows," the frank, kind look.

When the argument over the paper finally was settled, Rutherford told the young man who had come to him not long before "to get some experience" that he had proved unexpectedly obstinate. Bohr also was surprised by Rutherford. He had not expected the older man, who as usual was extremely busy with his research, to give so much time to the matter of the paper and to show, in Bohr's words, "an almost angelic patience." And Rutherford had agreed at last that except for trivial matters of language the paper had to be as Bohr had made it.

This was the beginning of a close friendship between the two men, of visits and a flow of letters back and forth between Denmark and England. Bohr has said that during their lifelong association Rutherford never spoke more angrily to him than he did one evening at a Royal Society Club dinner. He had overheard Bohr, conversing with some other members, refer to him by his title, Lord Rutherford, and in a rage, broke into the conversation to ask the Dane loudly and bitterly: "Do you lord me?"

Niels Bohr's quantum theory of the atom was published in three parts by an English journal during the year 1913. Some of the physicists who studied the theory felt that it was just "juggling with numbers until they can be got to fit." Not Albert Einstein, however. A colleague who told him of some new evidence in support of Bohr's theory said that as he listened, "the big eyes of Einstein looked still bigger."

"Then," Einstein said, "it is one of the greatest discoveries."

During the ten years after 1913 more and more physicists came to agree with Einstein, as experiment demonstrated that the theory could not be dismissed as number-juggling and as Bohr, having established the atomic basis of the science of spectroscopy, did the same for the science of chemistry. Now the vast quantity of information accumulated by these sciences could be interpreted in terms of the atom.

This decade between 1913 and 1923 was the early age of atomic physics and it was, one physicist said, "full of promise and despair"; for the very theory which made it possible to understand some things about the atom made it impossible to understand others. It was a time for young physicists. The data of chemistry and spectroscopy awaited interpretation and this could not be done on the basis of established science. A long training in classical physics, until the underlying ideas began to seem self-evidently true, was not as valuable as the ability to question established doctrine and entertain radically different ideas.

In the next chapter we will tell of some of the triumphs of Niels Bohr's theory and also of its failures during the decade when physicists were trying to understand a part of nature which defied logic and could not be represented by a model. Then we will go on to see how Bohr began to work with a group of young physicists, "living," one of them said, "as an equal in a group of young, optimistic, jocular people, approaching the riddles of nature with a spirit of attack, a spirit of freedom from conventional bonds, and a spirit of joy. . . ."

CHAPTER SEVEN

Niels Bohr: Early Days
of Atomic Physics

To those of us who were educated after light and reason had struck in the final formulation of quantum mechanics, the subtle problems and the adventurous atmosphere of these pre-quantum mechanics days, at once full of promise and despair, seem to take on an almost eerie quality. We could only wonder what it was like when to reach correct conclusions through reasonings that were manifestly inconsistent constituted the art of the profession.
—C. N. Yang, *Elementary Particles*, Princeton, 1961.

I N 1914, as the First World War began, two German physicists, James Franck and Gustav Hertz, announced the results of some experiments which had grave implications for Niels Bohr's new theory of the atom.

Before describing what happened, it will be useful to consider the kind of experiment which might be done to verify Bohr's idea that the energies which the electron, and therefore the atom, might have are restricted and that these discontinuous energies are reflected in the discontinuous line spectrum of the atom.

The basic equipment of the experiment would be a straight beam of electrons moving at controlled speed which is sent through a container filled with a gas of atoms. (Mercury gas could be used because its molecule consists of one atom.) After the electron-bullets leave the container their speed is measured to learn whether it remains the same. Any reduction in the speed indicates that the bullets have lost energy to the mercury atoms through which they passed.

According to Bohr's theory, the atom cannot accept *any* amount of energy but only the specific amounts which would send the electron from the first quantum orbit, representing the atom's normal or ground state, to one of the stipulated outer orbits. Thus the electron-bullets ought not to slow down at all unless their initial speed, or energy, is enough to send the target electron to an outer orbit.

Let us say that the energy amounts pertaining to each quantum orbit are 20 units of energy for orbit 2; 30 units for orbit 3, and so on. Suppose then that the experiment is repeated again and again, and each time the electron-bullets enter the container they are traveling at a slightly different speed. On the basis of theory one would expect to find a definite pattern in their end speeds: a bullet of less than 20 energy units should lose no speed, a bullet having energy between 20 and 30 units should lose no more than 20 of this, and so on. The last rung of this energy ladder would be reached when the bullet has sufficient energy to send one of mercury's electrons beyond the outermost orbit. At that point the neutral mercury atom, having lost an electron, would be positively charged. In other words, the atom would be ionized.

According to Bohr's theory, energy transferred to the atom by a bullet would be emitted as light having a particular wavelength as the electron falls back to an inner orbit. Thus it should be possible in such an experiment to observe the formation of the mercury spectrum virtually line by line as bullets of higher and higher energy are sent through the gas.

The experiment which James Franck and Gustav Hertz actually performed resembled this one. It was begun before Bohr's theory was published and Franck and Hertz were not acquainted with Bohr's ideas. They were interested in measuring the energy required to ionize the mercury atom. They were not looking for evidence of discontinuous atomic energies; they did not suspect their existence.

Since Franck and Hertz wished to determine at what point in the bombardment of increasing energies the mercury atom lost an electron and became a positively charged ion, the experiment was arranged so that they could detect a positive current of electricity in the container of mercury gas and thus learn how much energy was required to produce that current. And in 1914 they published their findings. They had succeeded in measuring a positive current; they assumed that the atoms had been ionized and announced that an energy of so much was required to ionize mercury.

This was bad news for Niels Bohr's theory: the mercury atoms had been ionized, said Franck and Hertz, the atoms had lost electrons. But before the electron was lost there should be—on Bohr's theory—a series of energy jumps and the Franck-Hertz experiment should have detected this stepwise energy transfer.

It had not been detected.

Bohr studied the report of the experiment. It looked bad for his theory. It looked bad *if* the mercury atoms indeed had been ionized. But suppose they had not been. Was there something else which could have produced the positive current Franck and Hertz had measured? There was. During the bombardment process the mercury atoms had emitted high-frequency light, light which upon striking certain metals has a known effect, the photoelectric effect. Metal electrodes had been used in the experiment. Due to the photoelectric effect, electrons could have been released from these electrodes. Perhaps this side effect of the experiment was responsible for the positive current, the electrodes having been, so to speak, ionized and not the mercury atoms. Perhaps the energies of the Franck-Hertz bullets had been sufficient only to send the electrons to the first quantum orbit and the energy figure they had announced as that required to ionize mercury was in fact the energy corresponding to the first rung of the energy ladder postulated by his theory.

Bohr was back in Manchester, where Rutherford had invited him to teach, when he made these calculations and he told Rutherford what he suspected about the Franck-Hertz experiment. The New Zealander replied to the effect: "Why don't you check this yourself?"

And so the construction of an intricate quartz apparatus with various electrodes and grids began. Bohr had help with it from

Niels Bohr in 1923. He was thirty-eight years old and the year before had won the Nobel prize for his 1913 theory of atomic structure. The photograph was taken at Columbia University during Bohr's first trip to the United States.

another member of the physics staff, and most important from an expert technician and glassblower who had assisted Rutherford for many years in the Manchester laboratory. (He had made the fine glass tubes for the experiment which identified the alpha particle as helium.)

At this time England was at war with Germany and the glassblower was a German national. Rutherford had asked his government to make an exception in the case of his glassblower and allow the man to remain in England despite the war. But he had a hot temper and his violent pro-German outbursts led to his internment by the British authorities.

At the time when that happened, Bohr's apparatus was finished. Before the experiment was completed, however, there was an accident. The support on which the setup rested caught fire. ". . . Our fine apparatus," said Bohr, "was ruined. . . ." And without the glassblower, he found its reconstruction impossible. That experiment never was done, but a few years later in New York other physicists proved that his suppositions were correct. Franck and Hertz had been mistaken in thinking that the mercury atoms had been ionized in their experiment. In truth they had found very strong evidence for Bohr's idea of discontinuous atomic energies and for this they were in 1925 awarded the Nobel prize. Niels Bohr won the prize in 1922.

It did not take physicists very long to recognize and to exploit what Bohr had proposed and experiment confirmed: the key to the atom's structure was to be found in the spectrum of its emitted light. This fresh knowledge indeed meant that there were more questions to ask.

Bohr addressed himself to one of them: "How do the properties of the different chemical elements arise from the different structures of their atoms?" Why does the presence of one electron more in the atom mean the difference between a liquid which forms many chemical compounds and a gas which forms none?

Since he first had gone to England to study under J. J. Thomson, Bohr had been interested in this problem. It was quite clear that a relationship must exist between atomic structure and properties of the various elements. These could be listed according to their mass—hydrogen first, then the slightly heavier helium and so forth—to disclose a repetitious pattern. The third element in the list, the 11th, 19th, 37th and 55th distinctly resembled one another. They were the soft alkali metals. Another striking resemblance was between the 2nd, 10th, 18th, 36th, and 54th elements in the list, all gases which do not combine easily with other elements, the so-called "inert gases." Such periodic repetitions showed up very clearly in the table of elements worked out by the Russian D. I. Mendeleyev, and the question arose: "Can the repetitions be explained on the basis of a repetitious arrangement of electrons within the atom?"

Before Bohr's stay at the Cavendish Laboratory, J. J. Thomson had been working on just this problem. His research was based on his own atomic model, in which positive electricity rather than being concentrated at the core was distributed evenly throughout the atom. Within this electrical field Thomson pictured the electrons as being arranged in closed rings—more precisely, shells, since the atom was conceived as a sphere. One shell of electrons was enclosed within another, like the skins of an onion. On the basis of this idea, J.J. and his students had been able to account for some of the observations of chemistry. Their success was one of the main reasons why prior to Rutherford's scattering experiments J.J.'s model was highly regarded.

The discovery of the nucleus changed this, but Bohr thought that J.J.'s shell idea still could be used and during the early 1920's he worked out a similar repetitious arrangement of electrons for the hydrogen atom, the helium atom and so forth, based on studies of their different line spectra as interpreted by his theory.

Bohr began by postulating a nucleus of a given positive charge, which was not surrounded by any electrons. Then, one

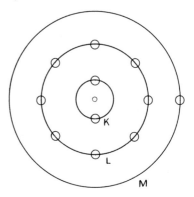

Electron shells of the sodium atom.

by one, he added electrons to the atom until there was a sufficient number to offset the positive charge of the nucleus. Each
electron was assigned to a quantum orbit; a group of orbits
formed a shell.

The first, or K, shell was the one nearest the nucleus and therefore most tightly bound by its positive field; the next shell, en-
closing K, was less tightly bound and so on. For example, the
neutral sodium atom has eleven electrons. Two of these would
form a closed shell, since according to Bohr every atom having
two or more electrons has such a K shell. The next shell, enclos-
ing K, is complete when it contains eight electrons and eight of
sodium's electrons would go into this L shell. One of sodium's
eleven electrons remains. The other shells are complete; the re-
maining electron starts a new shell, M, which is not complete until it also contains eight electrons.

This single, loosely bound electron accounts, Bohr said, for the
chemical properties of sodium. Why is the element "strongly
electropositive," as the chemists say? Because the single electron
in the uncompleted shell is lost easily, leaving the atom positively charged. Again in the terminology of chemistry, why is
sodium "monovalent"? Because due again to the "extra" electron,

the sodium atom tends to combine with atoms which lack one electron in their outermost shell, for example with chlorine atoms, forming the compound sodium chloride, or table salt.

Sodium is one of the soft alkali metals which, as mentioned earlier, have similar properties: all are monovalent and strongly electropositive. Again the shell structure explains why this is so: each element in the group has a different number of electrons but in every case, when the electrons are arranged in the shells which according to Bohr are possible, there is one electron left over, one easily lost unit in the outer region of the atom.

The resemblance among the inert gases such as helium and neon could be explained in the same fashion. Each of these elements has a total number of electrons which fit exactly into the specified shells, so that in the outermost shell there are no vacant places. Therefore these atoms do not as a rule combine with others. Chemically speaking, they are "inert."

This was the way Bohr explained the main features of the periodic table of elements on the basis of a periodic arrangement of electrons within the atom. In his book *The Search*, C. P. Snow, novelist and physicist, has described the feelings of a physics student when in college he first learns of this work:

> For the first time I saw a medley of haphazard facts fall into line and order. All the jumbles and recipes and Hotchpotch of the inorganic chemistry of my boyhood seemed to fit themselves into the scheme before my eyes—as though one were standing beside a jungle and it suddenly transformed itself into a Dutch garden. 'But it's true,' I said to myself. 'It's very beautiful. And it's true.'

Such were the victories of the Rutherford-Bohr atom. Yet for all the beauty and truth of Bohr's deductions, the theory on which they rested was inadequate, and before very long it would be replaced by a new and different system of ideas. Bohr's was not the theory which accounted both logically and precisely for the immense quantity of data accumulated by the sciences which studied matter from different points of view. It was not the theory which gave full mathematical mastery of the atom.

Certainly Bohr's theory pointed in that direction. The trouble was that it did not go all the way. Attempts to calculate from his assumptions the exact spectra of elements other than hydrogen failed. Although theory predicted the observed spectral lines, it also predicted others, frequencies which do *not* appear in the actual spectrum. The shell systems based on the predictions were not adequate to the task of giving known chemical data with invariable precision. It is true that Bohr could get a right answer by using his correspondence technique (that is, by calculating the frequencies which should occur, according to classical physics, and then using these classical predictions to "correct" those of quantum theory). This juggling of opposites, however, was closer to art than to science. Bohr could not proceed logically from theoretical assumption to precise fact.

What is more, he had no reason for stipulating, "Only two electrons to the innermost shell, only eight to the next," and so forth. When the electrons were divided up in this way, all the rest followed. But why should they be divided just so? When the atom was neutral, this meant, according to his theory, that all the electrons should have fallen from the permitted outer orbits to those which were nearest the nucleus and represented the atom's normal or ground state. Then why were not all sodium's electrons in the innermost shell, instead of just two?

In the next chapter we will see how someone found a grounding for the shell stipulations of Bohr. But even that great step forward did not eliminate the misunderstanding which lay at the very root of Bohr's theory. Today the strengths and weaknesses of that theory are understood very well. Bohr's idea that the atom's energies are discontinuous (quantized) was correct, but his explanation for this was not. The electron is not, as Bohr assumed, a miniature version of the material particles with which preatomic physics had been concerned. The electron does not move within the atom as such a particle would move.

It was a long time before this was recognized. It was a long time before someone asked the right question: "Can it be that just as light at times behaves like a collection of independent particles and at times like a continuous wave so do the elementary units of matter?" That question was asked only after more experimental evidence was found to support Einstein's quantum theory of light. In the beginning there was nothing but the single experiment on the photoelectric effect. The later and much more detailed experiments—done by the Americans R. A. Millikan and A. H. Compton—persuaded many physicists that light did indeed have properties which logically speaking contradicted each other. Until then physicists tended to think in terms of a choice between the wave and particle models and tried to fit information about light to the first model, information about the atom to the second.

They continued applying to the case of the atom those laws which account for the motion of material bodies from microscopic specks to planets, the laws of Newton, otherwise known as "classical mechanics." ° As we have seen, Bohr found that in order to match theory to the hydrogen spectrum the classical laws had to be restricted by means of Planck's constant, making energy discontinuous. Like Planck, Bohr had *corrected* the classical laws. When, later, it became clear that a corrected physics could not account for the atom (or for light) and that a wholly new system of ideas was necessary, physicists could look back at the work of Bohr and Planck and recognize that the fixed number

---

° Physicists also applied the motion laws of Einstein's special theory of relativity because the motion of electrons in the atom approaches the speed of light. In the present context Einstein laws may be considered classical in that they are a refinement of the laws governing motion of material bodies.

represented by h was a signal, like another constant—c, the speed of light. Einstein had shown that the c constant signified a limit to classical mechanics. Bodies which moved at speeds approaching that of light were governed not by classical but by relativistic laws, laws based on altogether different assumptions. The h constant signified that in the case of the atom classical mechanics again did not apply. Before the atom could be accounted for fully and logically, before it could be understood, there first had to be a mechanics based on assumptions different from those of Newton (and different also from Einstein's)— there had to be a quantum mechanics.

That would be a precision tool. While Bohr could account for spectroscopic and chemical data only by an artful juggling of contradictions, this data would flow logically, inevitably, from the assumptions of quantum mechanics. The physicist would be able to put some information about the iron atom into the quantum-mechanics machine, press buttons, so to speak, and out would come all the detailed properties of iron under every conceivable condition, given exactly. With quantum mechanics the unknown, like the known, could be deduced with precision. It was the theory which meant full mathematical mastery of the atom.

But in the dark age of atomic physics, the "pre-quantum-mechanics days," every mathematical victory raised a question which could not be answered. There were many quantum corrections to classical physics besides the ones made by Planck and Bohr and all posed the same sort of question: "What lies behind this limiting quantum rule?" "What causes it to be?" Thus in the case of Bohr's rule there immediately arose such questions as "How is the electron able to pick out a particular orbit in which to circle the nucleus?" "What keeps its energy from falling below a certain point so that it does not plunge into the nucleus?" In answer to these questions, theory gave not a force but numbers (among them the number represented by h). Not only was there no reason for the quantum rules, they seemed to suggest that the electron possessed astounding properties. For example (going back to the figure on p. 101), Bohr's theory said that an electron headed for orbit 5 would vibrate at a different frequency than it would if it were headed for orbit 4 or 3 or 2 or 1. In each case the electron's behavior at the start of a jump had to be geared to its goal. How could this be? It was as if the electron *knew* beforehand where it was going to jump; as if it *decided* what frequency at which to vibrate.

Each quantum rule raised a question of this kind. Each rule made it harder to visualize the workings of the atom. While one could diagram the Bohr atom, as we have shown, these representations were quite different from the models one may see on display in a science museum, the miniature motors which show how stresses and strains, pushes and pulls cause the motor to operate as it does. There could be no such model of the Bohr atom; there was no mechanism which explained how it worked. As the mathematical description of atomic happenings became more and more precise, it became simultaneously harder and harder to picture these happenings until at last, in the mid-1920's, the new atomic theory of quantum mechanics was born, the theory which at last answered the questions. Until then, physicists lived with them (and so shall we, but only for a chapter or so).

Could nature, they wondered, possibly be as "absurd" as their equations indicated? a striking example being the equation $E = hf$, which applied in the atom as it applied in the case of light, and which equated the broken and the unbroken, the limited and the unlimited.

The American physicist R. A. Millikan has said that he learned of Einstein's challenge to the wave theory of light—the theory of the photon—when he was working in a laboratory (under A. A. Michelson at the University of Chicago), "working . . . contin-

uously and familiarly with the wavelengths of light. . . ."
These were, he said, just as real to him as "foot rules and spring
balances." He called Einstein's idea that light had a particle-like
structure "reckless." (And then spent the next ten years of his life
on an experimental test of the reckless idea, at the end of which
time he had verified it in detail.)

During the same period Max Planck was arguing against the
theory. When he wrote to the Royal Prussian Academy of Science to propose Einstein as a member he said, "That he may
sometimes have missed the target . . . as, for example, in his hypothesis of light quanta, cannot really be held against him."

Einstein himself worried the problem of light's structure in an
office in Prague (this was shortly before he returned to Germany), an office which chanced to overlook a beautiful park be-
longing to a neighboring mental hospital. Gazing out of his window, he would watch the park inhabitants. Some, strolling back
and forth under the big trees, were lost apparently in deep
thought; others, gathered in groups, were—to judge from their
violent gestures—having heated arguments. Einstein, gazing at
the scene, was reminded of the behavior of his colleagues.
"Look," he told one, pointing out the view, "those are the mad-
men who do not occupy themselves with the quantum theory."

In 1917, Einstein made another major contribution to the
maddening theory. He found a route consistent with Bohr's
theory of atomic structure which led to Planck's radiation formula. The latter had been based on very general assumptions
about the structure of matter from which radiation arose, for the
electron was a new discovery at the time Planck worked out his
formula. Now Bohr had provided a detailed theory of that structure and it ought to be possible to deduce Planck's work from
Bohr's: specifically, it ought to be possible to deduce the energy
distribution of black-body radiation from Bohr's rule relating
frequency to the orbital jumps which were possible. (The num-

ber of times a jump was repeated would give the energy intensity of each frequency in the spectrum.)

Einstein succeeded in doing this and in so doing he demonstrated something of profound importance: one could go from Bohr to Planck by making statistical assumptions as to the occurrence of a jump. Following rules giving the probability that a jump would occur, then and only then could one deduce the spectrum of the black body. Please note, said Einstein in his paper, the resemblance between this case and the Rutherford-Soddy laws for radioactive decay. In both instances we do not know what causes a particular atom to act in the way it does. We do not know what triggers the electron's jump to one particular orbit rather than to another; we do not know what triggers the nuclear change which is responsible for radioactive decay. Therefore we may predict only by considering a vast number of like cases and so learn what it is most probable to expect.

Niels Bohr pounced upon this work of Einstein's, seeing in it the way to correct his atomic theory and make progress toward a new one. As noted, his theory did not—except in the case of hydrogen—yield the line spectra correctly; it predicted lines which were not observed as well as those which were. With the help of correspondence, Einstein's idea could be used to correct this weakness in theory. Thus, as mentioned earlier, Bohr would take his clues from classical physics as to which spectral lines were the most probable to expect. Knowing this, he could use statistical methods, as Einstein had done, and adjust his theory of atomic structure to the spectral evidence of that structure. These calculations made it possible to work out shell structures for the various elements, as described earlier.

Bohr was going ahead with such theoretical problems even though he knew very well that the theory he used was imperfect. He was going ahead despite the multitude of unanswered questions. As he saw it, there was no need to wait for experiments to be done which might answer these questions. The experiments *already had been done* by the spectroscopists. Not only had they accumulated data about the ordinary line spectra of the different elements but also about changes in a spectrum caused by different physical conditions, such as the presence of an electric or a magnetic field. These variations in a spectrum as the atom responded to different conditions constituted a huge catalogue of atomic behavior. The catalogue could not as yet be read: it was written in code, a code of light frequencies—the spectrum. If that code could be broken, by trying in case after case to find the matching mathematical expression, might this not lead one to the desired new theory, the theory which would answer the questions? Bohr believed that it might and he continued to work with imperfect theory, with correspondence and with statistics. Indeed he regarded the new application of probability laws, introduced by Einstein, as an embodiment of his correspondence idea: the statistical method constituted a path one could follow between old and new; it provided a language in which a final understanding of the atom could be expressed, an understanding which resolved conflicting ideas.

Einstein thought otherwise. The method of statistics appeared to him as something temporary. One used these rules, as he had done, when information was lacking. Once that information was gained, once physicists learned what prompted the atom's mystifying behavior, the need for statistics would be gone. While Bohr could conceive that statistics might play an essential part in a final understanding of the atom, Einstein could not. He took a different line of approach to questions raised by the quantum theory. It was the same line he had followed in his studies of black-body radiation and the photoelectric effect. Thus he investigated the properties of the particles of a gas, comparing these with the properties of radiation, seeking, as before, a resolution of the duality of continuous and discontinuous, of wave and particle.

Meanwhile Niels Bohr continued to pursue his own line of attack, following a statistical path through unknown territory, sniffing out clues and gradually, very gradually, beginning to sense the lay of the land. He learned the kind of solution to expect in various sorts of cases; he sensed the direction quantum theory was taking; he began to acquire a new way of looking at the unsolved problems; he began to ask the right questions.

While some physicists despaired of ever being able to understand the atom, Bohr on the whole remained optimistic, believing as before that what appeared contradictory would in the end prove to be part of a larger harmony. Strange as the new quantum ideas were, much as they denied man's former understanding of nature, there was between them and that older understanding a correspondence. One could take the physics which summed up that understanding, classical physics, into the quantum world of the atom and it enabled one to move ahead. Answers to the questions would come; understanding of the atom was possible.

Bohr was by no means the only physicist engaged in the work of matching theory to spectra. After the success of the Franck-Hertz experiment, interpretation of spectra became the major work of theoretical physics. It was done not only in Copenhagen but at the University of Leiden, in the Netherlands; and at the German universities of Munich, Göttingen, Tübingen, Berlin. To these universities came graduate students from other European countries, from Great Britain, even from Asia and from the United States. In striking contrast to the situation today, there was virtually no institution in this country where atomic research was done. Indeed there were only a few theoretical physicists here and Robert Oppenheimer has said that when he was a student at Harvard in the 1920's, he did not know that it was possible to *be* such a thing as a theoretical physicist. He learned how in Europe, as did Fermi, Szilard, Teller and many others who, since the atomic bomb, have become so well known.

A good number of today's theoretical physicists were attracted as young men to Denmark where Bohr was so successfully wielding what has been called "a magic stick which worked only in Copenhagen," his correspondence technique. The first to seek Bohr out was a young Dutchman named Hans Kramers who, it is said, learned to speak Danish during a trip between the Netherlands and Denmark. Kramers was just twenty years old when he first came to Copenhagen in 1916 to attend a student conference. When visiting a foreign city, he thought, one ought to look up others in the same line of work and since he was studying to become a physicist he went over to the University of Copenhagen to see who was in the physics department. There he found Niels Bohr, who had left Manchester not long before and was teaching in Copenhagen again.

The two young men liked each other at once and when at the end of his visit Kramers ran out of money, he thought of Bohr and looked him up again, this time to ask for a loan.

Bohr had been wishing that there were more physicists at the University of Copenhagen for him to talk to (or work with; it was the same thing), and so he asked Kramers if he would like to become his assistant. Kramers agreed; the university consented. During the next few years, Bohr invited some other young physicists—a Norwegian, a Swede, a Hungarian—to join him in Copenhagen and persuaded the university to hire them. He found some vacant rooms in an old high-school building near the campus where they could work together.

This was the beginning of a research center which would be called "The Institute for Theoretical Physics" but is known to physicists simply as "Bohr's institute." In 1920, when some Danish businessmen donated funds, the institute acquired its first building and became official. Since then more buildings have been added and the number of research students has grown from ten or so to almost one hundred. But the institute has remained, as it was to begin with, international and informal.

Since studying at Manchester, Bohr had hoped someday to have a school like Rutherford's. He liked the way the New Zealander included his students in the ideas behind the research and took their suggestions seriously. He even, in a way, liked Rutherford's bad temper. "He knows," said Bohr, "how to use his temper, which he does not try to hide, to a sound criticism of his own efforts and those of his assistants."

Bohr was a calmer person than Rutherford but he also let his moods work for him and was exceptionally direct in his relationships with students. Perhaps even more than Rutherford, he believed in collaboration between theorist and experimenter, between professor and student. The fresh viewpoints of the young were valuable, he thought, and so was their critical attitude which forced the teacher to reargue what he knew and understand it from other angles.

In the history of quantum physics Bohr and his students played a decisive part. Although Copenhagen was not the only place where professor and student together worked on what we have called the "code" of spectra, it was at Bohr's institute that physicists learned the correspondence technique by means of which that code was broken. It was there, in talk with Bohr, that they began to ask the right questions, the questions which would in the end enable them to understand.

Exploratory talk, argument, is for almost every theoretical physicist a necessity of life. In physics, particularly modern physics, the mathematical expression is apt to come first, its full implications only being understood afterward. Again and again the physicist is confronted with the kind of problem Max Planck had with his first formula for black-body radiation: How is the abstract logic to be interpreted? It may be his own formula, it may be another's; the problem is the same. And when he finds it impossible by himself to make out the meaning, he is apt to look for another physicist, who may be just as much in the dark as he is. Then, in turn, one will explain (argue) to the other

what he thinks the mathematical expression might mean. Often
the ordering of thought, in the attempt to present a logical case
—or to refute the other person's—will bring enlightenment.
Robert Oppenheimer has called this sort of talk "explaining to
each other what we don't know."

It is this need for argument, as well as their need for speedily
communicated information, which brings theoretical physicists
together often, informally as well as in conferences national and
international, small and large. For the sake of an argument the
theoretical physicist will travel (or telephone) great distances.
And in the 1920's there was no better place to find an argument
than Bohr's institute, where the barriers between professor and
student, which Planck had encountered at the University of Berlin, virtually did not exist and everyone was arguing with everyone else most of the time.

In the next chapter we will introduce two young men who
worked with Niels Bohr during the "eerie" and "adventurous"
period of atomic physics when rash speculation was the order of
the day. Their names are Wolfgang Pauli and Werner Heisenberg and their ideas played a fundamental part in the theory
which physicists worked out after the spectral code was broken.
The full name for this theory is "the Copenhagen Interpretation
of Quantum Mechanics." Despite their contributions, it is not
called the Bohr, the Pauli or the Heisenberg interpretation and
one reason for this is the fact that theoretical physicists talk to
each other so much. The interchange of ideas, the argument and
criticism necessary to their work, often makes it extremely difficult to label a particular discovery with one man's name. That is
one reason why the next chapter, before introducing Heisenberg
and Pauli, first will look in at Bohr's institute for a glimpse of life
among the group of young physicists who worked in Copenhagen during the 1920's and '30's.

CHAPTER EIGHT

Wolfgang Pauli, Werner Heisenberg, and Bohr's Institute

A happy era of physics that will not come again.
—H. B. G. Casimir

THE mystical atom-workshop"—that is what the young physics student, Fritz Kalckar, was expecting to enter as he trudged along Blegsdamvej Street for the first time, headed for Bohr's institute. Later he described what he found there and we quote from his account, warning the reader in advance that it is not to be taken too literally. (For example, the chief of the institute was not a dog.)

> A well-built body, shaggy grey hair, wise, melancholy little brown eyes under big brushed eyebrows, friendly, a little shy—that is the chief of the institute who already, far down Blegsdamvej Street, wishes one welcome. He leads us with friendly barking and friendly wagging tail up to the institute. . . . After repeated ringing of the bell and strong barking from Doggie, a shady-looking gentleman suddenly opens the door and asks crossly if it is someone with a bill, for then, "Come back another day." We assure the angry instructor that our purposes are friendly and he allows us to enter, where we disturb some young scholars, burning with curiosity, who are going through the day's mail, reading the postcards of everyone who isn't there and, if the paper is thin enough, also their letters.

The writer goes on to recount the next "great sensation," as from another room he hears a mysterious rattling sound and,

127

heart pounding, approaches nearer to see, with his own eyes, Niels Bohr at work with the atom. What is revealed in the next room is a "bombarding experiment" with many "pings" and "pops." It is being done not by Bohr but by two students, one of whom—the Russian George Gamow (now famous among other accomplishments for his popular science books)—explains that the object of the experiment is to hit the ball, for it was Ping-Pong, not atomic experiments, which had made the pop sounds.

As this account implies, anyone who wanted to see the work being done at Bohr's institute would experience difficulty. Except on blackboards, there was little trace of it. Few experiments were performed; most of the institute members were concerned primarily with theory—and at that time there were no formal classes.

On the outside, the institute—a three-story stucco building with a roof of red tile—looked as much like a residence as a school. Indeed, Niels Bohr, who had married in 1912 and in the ensuing years fathered a large family—five boys—lived in the institute for a while in an apartment on the top floor. Inside the building there was also a lunchroom, a library and various rooms where students lived and met for informal discussions. Bohr's sons, who often entered and left the institute through a back window, added to the informality of things.

It was hard, at first, to tell the blond boys apart. "These sons of Bohr's," said the physicist, Leon Rosenfeld, "were a great matter of speculation to me." Bohr's paternal air had impressed Rosenfeld at their first meeting—not a surprising impression, he said, in view of the fact that Bohr was surrounded by sons. The next day when Rosenfeld saw Bohr, again there were sons around him. "Different ones," thought Rosenfeld, and was bewildered when in the afternoon of the same day, Bohr appeared with what seemed to be still another boy. "He seemed," Rosenfeld said, "to stamp them from the ground or draw them forth from his sleeve, like a conjuror."

Mrs. Bohr, who was tall and beautiful, also added to the institute's domestic atmosphere; she liked to make sandwiches for the students who always were visiting the Bohr apartment, there to talk endlessly not only of physics but of politics and philosophy, music and books, movies and girls.

Besides Ping-Pong, which was played so much that there were two dips in the floor where the players stood, a favorite pastime was moviegoing. Niels Bohr, who admired a certain blond actress and all cowboy movies, often went along on these expeditions, one of which inspired what has been called "a little-known Bohr theory and an experiment to test it."

Bohr, with Gamow and some others, had seen a cowboy film and afterward—probably over beer and sandwiches (smørrebrød)—there had been an argument. "Why," said Gamow, "should the hero in these movies always draw his gun faster than the villain? After all, the hero is unprepared while the villain knows in advance what he is going to do. He would move more quickly."

Bohr, who always looked on the bright side of things, disagreed. Because the hero had not planned to kill, he would feel no guilt and therefore, Bohr thought, his reactions would be faster than the villain's.

As the argument progressed, it moved from the specific to the general, as discussions among theoretical physicists often do; and soon everyone was talking about a mythical country where the inhabitants owned and freely used pistols. In such a place, Bohr thought, the innocent would survive.

Gamow, however, was not persuaded. The next day, to settle the argument, he saw to it that he, Bohr and the others were armed with toy pistols. Later, at an unexpected moment, Gamow and his followers ambushed Bohr. Pistols were drawn. Victory went to Bohr, the innocent.

The movies inspired serious thinking, too. Rosenfeld, on his first trip there with institute people, was surprised to see a student, H. B. G. Casimir, start what appeared to be an important calculation before the lights went down and the movie began. "Poor Casimir," said Rosenfeld. "He had to wait until the lovers had safely got over their troubles and married and all, before he could resume his calculations. He did not lose a second either: every time the lamps lit up, they invariably disclosed our friend bent over odd bits of paper and feverishly filling them with intricate formulae. The way he made the best of a desperate situation was truly admirable."

Night was the time when many institute people did their best thinking. After the movies, the Ping-Pong, the café snacks, a student—alone in his room—worked hard, often staying up most of the night. He would sleep late the next morning, sometimes even missing that important event, the arrival of mail.

Those who were not asleep awaited it eagerly, not only for the personal messages it brought but the scientific news: the journals reporting recent experiments and the latest wild mathematical guesses. These journals were pounced upon before they went to the library, and when they contained something important word of it spread fast. A late riser, descending the stairs to the first floor, would know what had happened from the crowd of people gathered in the hall and the shouting, as everyone explained to everyone else what he did not understand.

After a while the crowd would break up as groups of two or three wandered off in search of a blackboard. While one person used it to develop his mathematical line of argument, the others would make themselves as comfortable as they could, sitting on a chair backward, sideways, knees up or lying down on a table with feet propped up against a wall. In this way, arguing in turn, they looked for the meaning of the latest mathematical news. If the mail had brought none that day, the same sort of discussion would occur, set off as a rule by someone saying, after having spent half the previous night trying to figure it out, that there was something which he did not understand. The exploratory arguments would continue until it was time for lunch, or some other meal. In Denmark, it is customary to eat five times a day. "Whenever you get a good idea," a student complained, "you have to eat."

Over the meal the arguers would relax and talk about non-scientific matters, such as the girls of Copenhagen. These girls possessed, it was said, "a wild and extraordinary charm." They also rode bicycles everywhere and this, according to a physical law discovered at the institute, explained why so many students married Danish girls. "When girls are riding by on bicycles," the law stated, "you see more per second." The young institute members also had formulated a system of classification into which girls could be fitted, though not, usually, without an argument:

1. You can't stop looking.
2. You can stop but it hurts.
3. It doesn't make any difference whether you look or not.
4. It hurts to look.
5. You couldn't look if you wanted to.

(The same classifications were used for movies, too; and if in consultation during a film, the group agreed that it was class five, they left the theater at once.)

Only when lunch was over would there be signs of formal study. Niels Bohr would ask a student to come to his office and tell him what he was working on. Encouraged by Bohr's questions, the student would develop his ideas, sometimes discovering in the process that they were faulty. Bohr himself never criticized.

Occasionally in the afternoon there would be a seminar. Perhaps a physicist from another city had come to visit Bohr, to argue a new idea; and the visitor would be asked to present his idea to the students as well. The speaker would begin, only to be interrupted by someone in the audience asking for clarification.

There was little shyness. As one past student puts it, "We knew each other well and felt quite free to say, 'I do not understand' or 'You are wrong.' "

Today the young men who studied with Niels Bohr in the 1920's and '30's are professors of physics and directors of research projects in Europe and the United States. Many of them serve as scientific advisors to the governments and military services of their various countries. They like to talk about the old days in Copenhagen. At times they laugh, recalling their extreme youth. "There was," one recalls, "a member of our group who made a really major discovery, on the order of Isaac Newton's. Why, it entirely changed our way of understanding nature. How did the youth regard his deep and profound achievement? Well, he reminded me of an avid stamp collector who had just gotten his hands on a rare item for the collection!"

Another physicist has mentioned the arrogance of those days, for many of Bohr's young men realized that history was being made in physics—and they were helping to make it. "We considered ourselves 'the select few,' " he says. "We were the ones 'in the know'; and like other arrogant people, we wanted to exclude others from our group. There was the feeling that students from certain countries could not possibly make good physicists and some of us even tried to keep student-applicants from those countries out of the institute. These self-appointed administrators thought that they were running the place. In some ways they were because in those days administrative matters were treated very casually. Bohr paid them little attention."

The physicists who speak here miss the old days even though they were, by their own admission, a little foolish and also more than a little poor. No one owned an automobile then; few could afford to ride first class on the train. A grant from the Rockefeller Foundation, which in the United States was considered no more than adequate for the grantee's support, in Europe seemed like a fortune. The student who won such a grant was wealthier for a while than his professors.

What is more, they expected to remain poor. At best, they could hope for a professorship, and these were few in the days when a physics department was two or three men. Competition for the scarce and far from lucrative positions was very keen.

Then why do they miss the old days? Now, by comparison, they are wealthy men. Also they are quite powerful. In the 1920's, no one outside physics cared what they were doing. But in today's atomic age their advice is sought by presidents, prime ministers and generals; their views are reported in the press.

But, they point out, all of this means that they are not as free now as they were in the old days. For one thing, when they were young money did not seem important. No one had much, and enough of it could be scraped together for the necessities of life, which included third-class railroad tickets to the various universities where other physicists were to be found. Old clothes were the mark of a physicist, whatever his rank; and they were worn with a certain pride.

The physicist, they say, was like a struggling artist because he found his work, unrecognized and unprofitable though it was, more exciting to do than anything else. No one chose his problems for him; he picked the ones which he thought were the most interesting. Not only was he free to choose his work, he was free to talk about it. He was never brought up short in the midst of a conversation with the sudden realization that what he was about to say was classified "secret."

In that faraway time before the atomic nucleus was split, a physicist's work was his own affair. It had no connection with military strategy, armament and the techniques of war. In those days a physicist felt perfectly free to follow his thoughts to their logical end. It did not matter where his thinking led him.

Among this league of carefree physicists, there were two who before they were twenty-five made outstanding contributions to the new science of the atom. Both men had been born just at the turn of the century when Planck's quantum theory appeared, and during the twenties of this century they also were in their twenties. Niels Bohr had met these two—Wolfgang Pauli and Werner Heisenberg—during a visit to Germany where he had gone to give a series of lectures. "Their extraordinary talent," he said, had impressed him deeply and he urged them to join him in Copenhagen. Perhaps they might cure the weaknesses of his own atomic theory.

In due course Heisenberg and Pauli arrived at the institute where they worked closely with Bohr. To this collaboration each one brought a different sort of ability.

Wolfgang Pauli, a portly young man from Vienna, resembled Buddha. His lips were rather full; his face, broad; his eyes, a bit slanted—but he was a Buddha with a gleam in his eye. At criticism Pauli excelled. The correct solution to a problem meant little to him unless the argument which led to it was terse yet complete, logically flawless. His scientific papers, few and far between, were end products of an arduous, lengthy thinking process during which the argument was worked and reworked until it met his exacting requirements. Rarely was Pauli's work wrong; his rigorous logical standards seemed to guide him always in the right direction. Other physicists marveled at what they called the "elegance" of his papers; frequently they sought Pauli's opinion of their own work.

In the role of critic, Pauli somewhat resembled the well-padded Mycroft Holmes who, rarely stirring from his comfortable armchair, solved problems for his more active brother Sherlock. But while many physicists profited from Pauli's criticism it was not easy criticism to take.

He offered to others what he wanted for himself. "There will be no heavy duties," he told the physicist who was to serve as his

A group at Bohr's institute about 1930. First row: Klein, Bohr, Heisenberg, Pauli, Gamow, Landau, Kramers. Second row: Waller, Hein, Peierls, Heitler, Bloch, Ehrenfest (Miss), Colby, Teller. Third row: first man unidentified, Wintner, Møller, Pihl, Hansen. Gamow probably is responsible for the toy cannon jocularly placed before Pauli, the tin horn before Heisenberg.

assistant. "Your job is, every time I say something, contradict me with the strongest arguments." * On the assumption that the people who sought his opinion felt the same way, Pauli looked at their work with a coldly critical eye. He questioned *everything*. He was ruthless, harsh, caustic and—very often—helpful. Heisenberg and Bohr were two who profited from Pauli's criticism, painful though it was. They admired his fierce intellectual honesty, Bohr comparing him to a rock in a turbulent sea. One always could depend on Pauli to say exactly what he thought.

And not only what he thought about your work. Once Pauli

* Another of this assistant's jobs was to police the plump physicist on his regular late afternoon journeys to the ice-cream parlor. There, as in the swimming pool which Pauli also liked to frequent, he was dutifully contradicted with the strongest arguments.

compared the sensitivities, the sore spots people have, to corns. In the long run, he said, the best way to get along with people is to step on their corns often, until they get used to it. And that is just what he did: deliberately, he stepped where it hurt, often causing misery to a new acquaintance. The pain did lessen, however, with repetition and with the knowledge that Pauli stepped on everyone's toes, the high and the low; that he did it, so to speak, out of principle. It was easier, for example, to be told by Pauli that you were hopelessly in error once you had heard him tell Bohr, "Shut up! You are being an idiot." "But Pauli . . ." Bohr would say, trying to continue his argument. "No. It's stupidity. I will not listen to another word."

Mathematical ability is apt to show itself early in life and Pauli was no exception to this rule. His talents had become apparent to his teachers when he was about eleven years old and they had allowed him to go ahead at his own rate and in his own way. Pauli had enjoyed school. At home, he talked about science with his father, a professor of biology, until he outdistanced his parent in mathematical knowledge; but he was closer, it is said, to his mother who shared his love of music. She was a journalist and during Pauli's childhood worked on a newspaper, while at home in her spare time she studied Latin and Greek. Her second and only other child—a daughter, Hertha—was born when Wolfgang already was a schoolboy and, since he was rather used to being the only child in the family, he did not regard her as an entirely blessed event. Hertha improved in his eyes, however, as she grew older and could be taught things. Eventually he became very fond of his redheaded sister, who loved to tease him.

The Paulis lived on the outskirts of Vienna, in a small villa surrounded by nut trees. The children gathered nuts together, swam in the Danube River and explored the Vienna woods. Sometimes Hertha's brother read to her from his large collection of Jules Verne novels, taking care to explain that in one of them,

From the Earth to the Moon, Verne had made a serious error. The passengers in a spaceship, Verne said, would experience weightlessness only when they reached the point in space where the moon's gravity compensated exactly for that of the earth. "Wrong," said Pauli. They would experience weightlessness as soon as the craft's propelling mechanism ceased to act and they escaped the friction of the earth's atmosphere.

On dark December afternoons, brother and sister would be told to leave the house so that secret Christmas preparations could be made. Though it was only about five o'clock in the afternoon the stars were visible. Young Pauli would point them out to his sister, passing on some of his large store of astronomical information. It was more than a strictly scientific interest he felt. The stars had for him deep personal meaning, as they had had for philosophers and mystics of ancient times.

Hertha was eight or nine, a little young for his astronomical knowledge. When he explained that the so-called "fixed stars" are by no means fixed in the firmament, she drew what seemed to her the logical conclusion: "So they fall."

Her brother was annoyed. He tried to correct her, but Hertha would not listen.

"They fall," she insisted at the top of her lungs. "They *fall!* They *fall!* They *fall!*"

This discussion ended, inevitably, in violence.

Grown up, Hertha became a writer, but before that she acted on the German stage for several years. This pleased Pauli, for he loved the theater; and he bragged about his sister to his friends. He liked to visit her after a performance, welcoming this opportunity to meet other actresses. (It is rumored that Pauli, who was decidedly a "night person," had one of his best ideas while watching a musical comedy.)

He was very young when he began to acquire a reputation as a physicist. His formal studies had been completed, a year ear-

Left: *After their initial encounter Pauli and Ehrenfest became friends. Here they are (Pauli on the left) riding the ferry boat to Copenhagen. In a moment Pauli is going to tell a joke.*

Right: *After the joke. The pictures were taken in 1929 (when Pauli was twenty-nine) by the physicist S. A. Goudsmit.*

lier than was usual, at the University of Munich, where he studied under Arnold Summerfeld, an outstanding theorist—the University of Munich had changed since Planck's student days. When Pauli was nineteen years old and a research student, a famous visitor had come to the university: Albert Einstein, to lecture on relativity. After Einstein had finished speaking, Pauli, in the audience, stood up. "You know," the teenager began, "what Mr. Einstein said is not so stupid . . ."

Actually Pauli at this time was well equipped to criticize Einstein. He had written an article on relativity for a scientific encyclopedia, an article which presented the theory with such conciseness, depth and logical beauty that Einstein, after reading it, said that he himself understood the theory better now.

As a result of this article on relativity, Pauli's name became known to other physicists; word also spread that the brilliant young man had a biting tongue.

"I do not mind," he told one physicist, "if you think slowly. But I do object when you publish more quickly than you think."

And when, at a conference, he was introduced to Professor Paul Ehrenfest of Leiden University, a very distinguished physicist whose papers were greatly admired by his fellows and who himself was an admirer of Pauli's relativity article, Pauli was very rude. Ehrenfest told him quite frankly, "I like your publications better than I like you."

To which the young Pauli crushingly replied, "Strange. My feeling about you is just the opposite."

This "terrible * young man," as he was sometimes called, arrived in Copenhagen when he was twenty-three years old. His reputation had preceded him, for Pauli's father had told someone from the institute that he hoped, when his son went to Copenhagen to learn physics from Bohr, he also would learn some good manners.

Pauli was gloomy and preoccupied when he arrived in Copenhagen. He was trying to understand something called "the anomalous Zeeman effect," referring to a change in line spectra in response to a magnetic field. The Bohr theory, even with its added rules, could not account for the observation that sometimes a single line in the spectrum splits into six or more when the element is placed between the poles of a magnet. Because this effect was inconsistent with theory, it was called "anomalous" ("Zeeman" is the name of the physicist who first noted the splitting effect which did fit theory). Pauli had been brooding about the problem for some time. When Mrs. Bohr asked him, in her motherly way, why he seemed so unhappy, he replied fiercely,

* Meaning "one who terrifies."

"Of course I am unhappy! I cannot understand the anomalous Zeeman effect."

It was possible, he thought, that the solution to this problem might also solve another—the serious logical gap in the shell system which Bohr had worked out to explain the periodic table of elements. That system, as we have mentioned, needed a backbone. By dividing up electrons, two to a shell, eight to the next and so on, it was possible to explain chemical properties, but what was behind this arbitrary arrangement of the electrons? What made them group in just this way?

Bohr had said something to this effect when he lectured in Germany; Pauli, in the audience, had received "a strong impression," he said later, that Bohr was looking for a general rule to justify his shell arrangement. Now, working in Copenhagen, Pauli discovered this rule for him, which is known as the "exclusion principle."

Pauli's work was based on observations of the spectra of atoms. As we have said earlier, the number of these observations was very great; in this immense heap of data, he distinguished a simple sorting-out principle which held in all cases. It was this: In any system of elementary particles, such as the collection of electrons within the atom, no two may move in the same way (occupy the same energy state).

Pauli's principle was general as well as simple, and in the years to come it would be found to apply in ways which no one suspected in the beginning: for instance, it governed the behavior of nuclear particles which were not discovered until years after Pauli put forward the rule. Indeed this work would form an important part of the quantum mechanics which lay in the future; it would remain a fundamental of atomic physics while Bohr's idea of how the electron moved in the atom would be discarded. Nevertheless, much of the work which Bohr and others based on this idea and corrected with correspondence was sound. It was not discarded but just understood differently. This was true of the shell systems for the different chemical elements which Bohr put together. And it was Pauli who found the reason for Bohr's assignment of electrons to the various shells. For the exclusion rule, when applied to Bohr's idea of atomic structure, restricted the behavior of electrons in just the way Bohr had restricted it, without a reason, in order to build his shell systems.

In return for this gift to Bohr's theory, Pauli received all the chemical data which the shells explained, to support his exclusion principle. This, added to spectroscopic observations—including the anomalous Zeeman effect—constituted overwhelming evidence that the exclusion principle was correct. But what did it mean? It did not offer a force which explained why, when an electron moved in a certain way, all other electrons in that system were prevented from doing the same. Like Bohr's atomic theory, it suggested that the electron possessed astounding properties, enabling it to "communicate" with its fellows and tell them how it was moving. Pauli's principle, which he presented in an elegant logical form, did not become meaningful until later, after the code of atomic spectra was entirely deciphered by means of a new mechanics. Until then, physicists called the principle, as formulated by Pauli, "baffling and beautiful."

When Pauli was working toward it in Copenhagen, he shared an office with one of the few experimental physicists at the institute, George von Hevesy, who had originated the radioactive tracer technique now used widely in biological research. At the time Pauli joined him, Hevesy was trying to identify a new element, predicted on the basis of Bohr's shell system. Already it had been named—"hafnium," after the Latin word for Copenhagen—but to prove its existence conclusively was a ticklish business because its spectrum closely resembled that of another element. After many weeks bent over his instruments, Hevesy was close to success; that was when Pauli began to use a desk in the same room. Unfortunately for Hevesy, the plump Viennese physicist had a nervous habit. When he was thinking, he rocked his body back and forth, and the harder he thought, the harder he rocked. The floor of the room was uneven, which made Hevesy's instruments shaky, and once in a while the vibration set up by Pauli's rocking would make it impossible to get an accurate reading. Then Hevesy would have to ask Pauli to stop.

Pauli did not know he was rocking. In fact, deep in his calculations, he barely was conscious that the room contained another occupant. Suddenly, Hevesy's request brought this fact to his attention and the fact that the occupant was doing work of some kind. That there were physicists in the world who performed experiments, Pauli knew, but only dimly. He was notably lacking himself in the talents necessary for experimentation and could not understand why anyone would enjoy this kind of science. Now that his train of thought had been interrupted, he asked Hevesy what he was working on.

"Hafnium," was the reply, and Pauli returned to his calculations.

Some days later, Hevesy again had to ask Pauli to stop his rocking so that he might read his instruments. Reminded once again of the existence of Hevesy (who, despite adverse circumstances, did, with D. Coster's help, succeed in identifying the new element) it occurred to Pauli to ask the other occupant what he was working on *now*.

"Hafnium," Hevesy replied again, probably between his teeth.

"I can hardly believe it," Pauli remarked, amazed that anything so inconsequential in his scheme of things could take so long.

The period when Hevesy was his roommate was one of the few times in his life when Pauli entered a laboratory. His total indifference to experimental science was well known, as was his clumsiness. He approached the simplest mechanism with deep suspicion, and it was rumored that before he could qualify for a driving license he first had to take one hundred lessons.

Summarizing various aspects of Pauli's work and personality —his great critical powers and their crushing effect, the mysterious forces implied by his exclusion principle, and his physical clumsiness—physicists began to speak of something called "the Pauli effect." When he was merely in the neighborhood, they said, this effect caused laboratory equipment to break or fall apart. There was the time, for instance, when an explosion—set off no one knew how—destroyed some vacuum equipment at the University of Göttingen. Now the cause was understood. Just at the time of the explosion a train transporting Pauli had stopped at the Göttingen station.

The Pauli effect, which became a legend of modern physics, could be understood as a compliment to its namesake. Theorists quite often are awkward in the laboratory, and ascribing such vast destructive powers to Pauli was one way of saying that, as a *theoretical* physicist, he was superb. In any case, Pauli himself was amused by the idea, particularly when at a conference in Italy some young physicists attempted to play a joke on him and demonstrate the Pauli effect. They had rigged up an intricate contraption involving a chandelier which was supposed to crash down in spectacular fashion when Pauli opened a certain door. But the "experiment" went wrong due to a rope wedged in one of the pulleys, and when Pauli opened the door nothing happened. Noting the elaborate apparatus which had failed its inventors, Pauli told them gleefully that they had succeeded in demonstrating a typical Pauli effect.

After a year in Copenhagen he had gone to teach at the University of Hamburg. This was in northern Germany, not far from Copenhagen. One took a short train ride to the coast and a ferry across the Baltic Sea. Pauli made the trip often. He was a familiar figure at the institute during the 1920's and criticism and suggestions for which he never received formal credit were behind many important papers. He was also among the first to recognize the limitations of models and to stress that in attempts to explain the atom they should be avoided.

One of those who profited from an association with Pauli was Werner Heisenberg, the other young man whom Bohr had met in Germany and invited to the institute. Heisenberg was handsome, blond, muscular. Like Bohr, he excelled at sports; often he wore *Lederhosen,* the leather shorts of a mountaineer. But at the institute his energies were invested mainly in physics. Most of the time he worked hard; rarely was he seen at the movies or the Ping-Pong table.

His companions knew him as a highly intuitive physicist. As Heisenberg himself put it, ". . . I must start not from detail but from . . . a feeling I have about the way things should be." In a flash he was able to see the solution which others failed to recognize despite a longer training and a greater knowledge of mathematics. But unlike Pauli, Heisenberg did not struggle long and hard to perfect the argument which led to his right answer. Unlike Bohr, he did not circle round and round a problem until he understood it from many different points of view. Heisenberg's tendency was to publish his findings as quickly as possible and let someone else worry about the argument. His mathematical proof was "crude," physicists said (Pauli used stronger words), and he had been known to come up with the right answer for the wrong reasons altogether. Pauli and Bohr urged him to develop the implications of the ideas which came to him so easily.

One of these ideas, as we shall see, enabled Heisenberg to break the code presented by the line spectra of atoms. He performed this feat by working only with atomic data which were related to observable fact. He made no assumptions based on a model. He did not use a model at all. Strangely enough, he had felt suspicious of atomic models long before he ever heard of

F. HUND

Werner Heisenberg at the age of twenty-four, not long after he deciphered the code presented by line spectra.

Niels Bohr's, as long before as the day when turning the pages of one of his first science textbooks he came to a diagram depicting atoms, and felt sure that it could not be right.

He was a schoolboy then, at the Maximilian Gymnasium of Munich, the same school Max Planck had attended forty years earlier. And, like Planck, Heisenberg there had enjoyed Latin and Greek and received a first and strong impression of what he called "real science," something altogether different from the engine building, the scientific play, which until that moment had absorbed him. He had been listening to his teacher present some axioms of geometry, feeling that it was "very dry stuff" when it suddenly struck him (as it had struck Einstein at about the same age) that there was something about geometry which was, in Heisenberg's words, "remarkably strange and exciting." This formal logic was not a thing apart, but actually fit the structures of one's own experience.

Stirred by this idea, Heisenberg began to play a new game. He tried to describe with the mathematics he knew what he saw in the world around him. He enjoyed the game and began to read books of mathematics to acquire more descriptive tools. In this way he learned calculus, with which he could formulate the laws governing the engines he had built.

He was learning physics. But it was the mathematical part of his game which interested him most and not the engines and other things which the mathematics described. This changed after he came upon the picture of atoms in his textbook. It was intended to show the atomic structure of a gas and some of the atoms were connected to each other by means of hooks and eyes, which represented their chemical bonds. "Idiotic," Heisenberg called those hooks and eyes. Why, right in the book it said that the atom, as named and defined by philosophers of ancient Greece, was the smallest indivisible building stone of matter. Logically, this meant that the atom had to be simple. It could not possess complicated properties because it was by definition something which explained such properties. Those things in the textbook illustration with a structure so elaborate that it could be represented with a hook-and-eye arrangement had no right, he felt, to the name "atom." He was angry that such an illustration was allowed to appear in a serious scientific book.

His instinct, he found out later, was correct. The atom which Democritus had proposed five hundred years before the birth of Christ had no physical properties such as color, smell and taste. It was something—a quite abstract something—which explained these properties. The scientists who came after Democritus used the word "atom" as he had used it to mean an elementary or ultimate particle. At length, they discovered that what they were calling the "atom" was composed of still other particles and therefore was not after all the smallest indivisible unit of matter. The electron, proton and neutron were much closer to the atoms of Democritus. Historically, the word had been applied to the wrong thing.

One of Heisenberg's friends, with whom he often hiked, was interested in the philosophers of ancient Greece and had read about their atomic ideas, some of which were even more abstract than those of Democritus, defining the atom as a mathematical form, a form without substance. This friend heartily seconded Heisenberg's opinion of the textbook illustration; indeed, he went even further. "The whole of modern physics is false," he declaimed, for in a book he had seen atomic models, based on Bohr's, which depicted the atom as an elaborate affair. There was no hope, he thought, for a physics which relied on a visual image of the atom, no matter *what* form this image might take. Heisenberg would not go that far. The models were wrong, he agreed, but perhaps there was something in the theory behind them. His curiosity was aroused; he wanted to know "the case for atomic physics."

He was seventeen then, and in his last year at the *Gymnasium*, an important year in his life. It was 1919, Germany had been defeated in the war, and in Munich workers and some of the returning soldiers had taken control of the city government by force, as the Bolshevik party recently had done in Russia. Heisenberg, with some other boys from his school, was serving in a military unit which was fighting the revolutionaries. At a very early age he had taken sides in a fierce political struggle.

For him it was a time of adventure. He still was attending classes but at any moment, day or night, he could be called away for military duty. He was exempt from the discipline of school, the discipline from which Einstein had rebelled. Neither parents nor teachers controlled what he did or where he went, and his military duties were not onerous. Often, particularly in the morning, he was free to do as he wished, and with a book chosen at random he would climb up to the roof of army headquarters, lie down in the sun and read. There he learned more about the ancient Greeks' atomic ideas, when he chanced to read Plato's dialogue *Timaeus*, which explained the infinite variety of nature on the basis of insubstantial geometrical forms and their combinations. He was impressed, as his friend had been, by Plato's reasoning. It was not based on experiment; strictly speaking, it could not be called "scientific." It seemed reasonable to him, however, that the fundamental units of matter which explained its infinite variations and properties would themselves be highly abstract—forms rather than substance—and he remained critical of all attempts to represent them visually.

Like Pauli and Planck, Heisenberg attended the University of Munich, where in the beginning he studied mathematics, for he still was more interested in that science than in physics. Then, when he was nineteen years old, he decided to pay a visit to Göttingen for what was called, jokingly, "the Bohr festival season," namely, lectures given by Bohr in Göttingen which drew a large audience of physicists and their students from many different universities. This was Heisenberg's opportunity to hear the case for atomic physics from its foremost practitioner and model-maker.

At Göttingen, the nineteen-year-old Heisenberg listened to Bohr's lecture on current notions of the atom and did not hesitate to point out to Professor Bohr, during the question period, where he thought the arguments were weak. Bohr answered this criticism but must have felt that his answers were inadequate because, after the question period was over, he went to Heisenberg and made him an invitation. They would go for a stroll through the countryside surrounding the old walled town of Göttingen and find a café, situated on high ground for a good view of the country. They would drink beer, have a snack, and talk some more about the matters Heisenberg had raised. We will have, said Bohr, "a good time."

On this occasion the two men spent several hours together and Bohr came away impressed by Heisenberg's talent. Heisenberg, as a result of this meeting, gave up his intention to become a mathematician and began to study physics. Bohr's way of doing that science impressed him, Heisenberg said later; he liked the way the Dane looked for ideas which might fit experimental findings rather than treating a problem mathematically from the start. The proof of the ideas, that is, their mathematical working out, came later, a necessary follow-up, but only *after* one had reached an understanding.

During this first long talk between the two, Bohr granted that the questions Heisenberg had raised earlier could not be satisfactorily answered as yet. Hearing the father of current atomic theory speak of serious unsolved problems inspired Heisenberg. Weaknesses in physics which he had recognized as a reader of that science, an outsider, suddenly came alive. These problems were real, immediate, urgent. Returning to Munich, he studied physics and a few years later, his formal education complete, he arrived in Copenhagen to work with Niels Bohr.

Heisenberg was, people said, a "radiant" young man, well endowed in body and brain and extremely self-confident. "The world was," one acquaintance said, "his oyster." And Heisenberg, at the age of twenty-three, was the one who discovered the key to the code of atomic spectra, thus founding quantum mechanics and demonstrating that Niels Bohr had not been following a false scent after all.

Heisenberg's achievement, great though it was, did not lead at once to an understanding of the atom. In the case of quantum theory, the mathematics came first and then the meaning. Between one and the other was a period of nearly two years when physicists still did not understand the significance of Planck's constant, when they still could not answer the questions raised by Bohr's theory and Pauli's exclusion principle and did not know why they had to resort to statistical rules.

Because of this fact, we interrupt our account of men and their work for a chapter which answers the questions which have been raised so far in this book, explaining what the physicists themselves did not know until several years after Heisenberg's work. This chapter of explanation will give the reader an advantage over the physicists and from that vantage point he can observe how they came to an understanding when we return to Werner Heisenberg and Niels Bohr.

The explanatory chapter which follows takes the form of a dialogue between two physicists, Oldfield and Newcomb. Real physicists, explaining their science to each other, would use technical language and a blackboard. They would make references to work which would not be familiar to a nonphysicist who was listening in and trying to learn something. Oldfield and Newcomb, the two physicists who speak in the next chapter, will not present these problems to the reader. They are imaginary.

CHAPTER NINE

An Introduction to Modern Quantum Theory

Now I know what the atom looks like!
—Lord Rutherford, 1911

Today we not only have no perfect model [of the atom] but we know that it is of no use to search for one. . . .
—Sir James Jeans, 1942

IN THE DIALOGUE between two imaginary physicists which follows, Oldfield speaks first. He knows nothing of developments in physics after 1924, being acquainted, then, with the Rutherford-Soddy theory of radioactivity; Planck's equation $E = hf$; Bohr's atomic model and the other work which has been discussed in this book. Having found his way out of the South American jungle where, conveniently for our purposes, he has been wandering from 1924 until now, he is curious about where this work has led and has come to Newcomb, a young physicist who can bring him up to date.

OLDFIELD: I have quite a few questions, Newcomb. First of all, what *is* light: portions of energy or a wave vibration? Second, do you understand the atom? How can the electron "pick out" a particular orbit in which to move and before it jumps to a different orbit "decide" the frequency at which it will vibrate? How can it "know" what the motions of other electrons in the

151

atom are? How can it "tell" them about its own motion? Bohr's theory and Pauli's principle gave no reason for this behavior, which is remarkable, to say the least. Bohr simply applied Planck's quantum constant to the case of the atom. He said that the atom is stable because of a limitation on the energies it may have, a limitation governed by Planck's constant. But you cannot call that number a reason! It does not explain how the electron is able to remain outside the nucleus. If you do have the answers to all these questions of mine, then you must have gotten rid of the statistical rules. Is that the case?

NEWCOMB: The answer to your first question is that light, when it travels, is a wave motion. When it strikes matter, however, light energy is transferred in units. And when matter emits light, again it is in units of energy.

OLDFIELD: In that case, there has been no progress since 1905, when Einstein explained the photoelectric effect with the idea of photons. The wave character of light was well established at that time, from interference experiments.

NEWCOMB: On the contrary, we know a great deal more. We know that the elementary units of matter possess wave properties; like light, matter has a double character, sometimes appearing wavelike in composition, sometimes particlelike. And again Planck's constant, the quantum constant, plays a fundamental role in this. I am sure that you recall the equation $E = hf$, where the symbol on the left side of the equation refers to a kind of particle; that on the right to a property of waves, with Planck's constant measuring the relationship between the two. Later, in connection with the study of matter, another and very similar equation was discovered. It is:

$$\text{momentum} = \frac{\text{Planck's constant}}{\text{wavelength}}$$

Again the left side refers to particles, since momentum is proportional to mass; and a wave property appears on the right. Here again, h measures the relationship between the two.

OLDFIELD: Then instead of one, you have two absurd equations! A wave of radiation is a state of tension of space. It has no boundary lines, no substance. A material particle, of course, is just the opposite. Yet the equations unite them.

NEWCOMB: Before going on with this, I think I ought to remind you of the size scale we are discussing. With our eyes, we are able to see something which is one thousand times smaller than ourselves. With technical aids, we can observe something even a million times smaller. But photons and elementary particles are one hundred million times smaller than our own bodies. We never will be able to see this region of nature; we never will be able to experience its events, which last one-millionth of a second or less.

Until quite recently we were unaware of this region. Then, from indirect evidence—such as scattering experiments—we learned of it and also learned gradually that it is not like the world we experience with our senses, the world of all our previous experiments. Atoms and photons are not simply smaller than what we know; they are different from what we know. This placed the physicist in a strange position. In an attempt to understand the new realm, he asked questions, questions like the ones you asked me. They are reasonable questions but they are based on experience in a different realm altogether and have little or no application to the subject at hand. And when the physicist performed experiments to answer his questions, naturally the answers he received did not make sense, just as my replies do not make sense to you.

OLDFIELD: Do you mean that the composition of matter and of light is neither particlelike nor wavelike? Do you mean that elementary particles and photons cannot be classified as either having a definite location or not having a definite location?

NEWCOMB: Yes.

OLDFIELD: That means the new region could not be represented by a model; it could not be pictured. All that you have to go by are indirect measurements and there is no way to make sense out of them. You try to form an idea of what they mean. Inevitably, your idea stems from the past experience of science; that is what you know. Inevitably, the idea is wrong. How could one possibly think of the right question to ask, so that the answer *would* make sense?

NEWCOMB: We have come to the core of the matter. The fact is that we do understand the atom. How? By learning when it made sense to ask a certain logical either-it-is-one-thing-or-the-other question and when it did not make sense to ask it. The science of quantum physics sets up limitations—guides—so that in our investigation of the new region we may continue to use ideas formed in the past and continue to use our logic.

OLDFIELD: In that case, it looks as if I must either become a student again and learn quantum physics or else remain curious.

NEWCOMB: No. I think that I can help you right now with your questions. We'll come back to them. First let's talk about an experiment which demonstrates why we require both a particle and a wave model to interpret the behavior of elementary particles such as the electron.

You are familiar with the apparatus which is used to demonstrate the wave character of light—basically, a source, an obstacle in which two slits have been cut, and on the other side of it a photographic plate or a screen which registers the arrival of the oncoming beam. Suppose we substitute for the light beam a beam of electrons, all moving at the same fairly low speed in the direction of the slits. The impact of each electron produces a tiny flash on the fluorescent screen. That is what one would expect to see on the assumption that the electron is a particle. After a while, however, as more and more electrons hit the screen, something else happens: the particles make a picture for us! A series of evenly spaced stripes appears. It is the same interference pattern which light makes under similar conditions. It resembles the water pattern which can be produced in a ripple tank. How can we explain this picture? A particle cannot steer itself to a certain area of the screen, and avoid others, to make a pattern.

OLDFIELD: And that is just what a wave can do, when it is split so that the two beams are precisely out of phase, some crests and troughs reinforcing each other, some canceling each other out. I see the problem: the particle model of the electron cannot explain its ability to interfere with itself and produce a pattern. But the wave model, which does account for the pattern, fails to explain the impact points, the flashes out of which the pattern is made. If we stay with the wave model, we would have to conclude that the extended rays coming from the slits suddenly contract to a point before hitting the screen. A self-contracting wave is as bad as a self-steering particle!

NEWCOMB: You agree, then, that both the wave and the particle models are required to interpret this experiment?

OLDFIELD: Wait! What happens if a recording instrument is attached to each one of the slits? The instruments will respond only to particles. Have you done that?

NEWCOMB: Yes. The instruments respond.

OLDFIELD: Then the electron is a particle when it passes through a slit and it is a particle when it hits the screen.

NEWCOMB: But under those conditions an interference pattern does not appear on the screen. Therefore the electron has not displayed its wave properties. This is the sort of thing which occurs invariably when one attempts to pin down the electron as either a particle or a wave vibration. In order to detect a particle on this size scale, an instrument which acts as a target must be used: a counter, a scintillation screen, a cloud chamber; and the wave properties of elementary particles are displayed only when there is no intercepting target.

OLDFIELD: Perhaps your experiments are not good enough. One of you young men will find a way to get around this sooner or later.

NEWCOMB: There is no reason to get around this! You are assuming that the electron really belongs in one of the two categories, particle or wave. But the evidence that it does not belong in either one is overwhelming: I just gave you a tiny piece of it. You say that the two things are contradictory? Contradictory to what? What we know. What do we know? Objects and forces on our own size scale. When you think about it, isn't it rather likely that the things which underlie the familiar and explain why it is the way it is would not be the same, would be unfamiliar? That is the way it has turned out. The so-called contradictions of the atomic region have helped us to understand familiar things.

I want to point out something else to you and so, purely for the sake of argument, let us assume that the electron really *is* like phenomena on our size scale and therefore must fall into either the wave or the particle categories, as does everything else on our scale. So we will begin with the false assumption that the electron either is or is not a particle. How could you find out?

OLDFIELD: To prove that it is a particle one would have to fix the electron's exact position in space and its exact mass. Such measurements could only be made indirectly, by observing the effect of something which came into contact with the electron. The least possible disturbance is essential, so the contacting object could be no more massive than the electron itself. That limits the possibilities: the lightest unit of matter *is* an electron. Suppose we shoot a beam of electrons through a target of atoms to a scintillation screen and from the scattering pattern . . . It wouldn't work. One must know the exact position and velocity of the bullet-electron at the time it contacts the target-electron in order to gauge the effect of the collision and from that arrive at an exact measurement. And of course you do not know it. That is what you are trying to find out in the experiment!

But there is another possibility. Matter could not be used; radiation could be.

NEWCOMB: No. The problem would be the same. When light interacts with matter it behaves like a particle. A photon-bullet would change the position and velocity of the target just as would an electron-bullet. To gauge the change, again one must know the exact velocity and position of the photon at the time of the collision, and there is no way to know this.

OLDFIELD: Therefore even if the electron were an ordinary particle, it could never be proved. But, as you say, we know from other experiments that the electron is not an ordinary particle: it has wave properties. So the fact that we cannot prove its particle identity is not bad news.

But I see something else which is. *Any* measuring tool which could be devised now or in the future must consist of matter or of radiation. There *is* nothing else. It follows that every measurement of the atomic region is and must be inconclusive. The very tools one uses to take measurements cannot be precisely measured themselves. Therefore it is impossible to allow for their exact effect on the atomic subject matter. One cannot separate the effect of what one does in the experiment from the subject of the experiment and know which is which. That *is* bad news.

NEWCOMB: And that is why I asked you to argue on the basis of a false assumption. You are correct. When the subject of an experiment is atom-sized, it is not possible to distinguish between it and the effect of the measuring tool. What do we do? Treat the two as one and go on from there.

OLDFIELD: In one way, this is nothing new. The subject of any experiment, atomic or not, must respond in some fashion to the tool. When you take the temperature of a hot liquid, some of the heat energy one is attempting to measure will be used to drive up the mercury column in the thermometer. One allows for that. Often it is impossible to allow exactly and then the result is stated as plus-or-minus so much. Within that range, however, the result is definite. It refers to the process or object one is measuring, to the thing itself and not what you have done to it. Always in science there has been a distinction between the two. When this distinction cannot be made, it means a radical change in scientific method.

NEWCOMB: Which method, let me remind you, was developed in studying rather large objects and gross changes in energy. When these are the subject matter, the scientist is able to keep track of the changes in nature which he causes by observing them. He uses light to observe and when he uses it to look at a planet, a gnat, even a living cell, he does not have to worry about photon collisions. It is easy to see how we formed the illusion that it would be possible always to control the effects of our observation. Now the illusion is gone.

Let me point out, however, that when an illusion is lost, it usually means that one has learned something. In this case, we have learned about the relationship between nature and the observer —who is, after all, part of nature himself. In physics now we are studying the ultimate components of all matter, living and non-living. Biology and physics are not as separate as they used to be. It seems inevitable to me that in penetrating more and more deeply into matter, we would come to the point where it was possible no longer to pretend that we are some kind of bloodless, sterile observer. I see it as a sign of progress that in physics, the most exact of the natural sciences, we have acknowledged in precise mathematical terms that this is not the case. I believe it hints that someday we will be able to understand living matter—ourselves.

OLDFIELD: Do other scientists feel this way? Surely not everyone is pleased with the idea that his results are not as objective as in the past.

NEWCOMB: How one regards this development in physics is a matter of personal philosophy; I was expressing my own. It is true that some scientists are unhappy about this aspect of quantum physics, which does represent a great departure from the past. Most often it seems to be the older men who feel this; the younger ones don't worry about it. The main thing for them is "Find out, no matter how." And they do. The fact that we cannot separate out the effect of our observation has not prevented us from reaching a very good understanding of electrons and of the atom as a whole.

OLDFIELD: How on earth have you done it? In the first place, you have shown me that elementary particles cannot be pictured. They combine the properties of definite location in space and no definite location in space; of substance and no substance. Not only is it impossible to visualize such things; it is impossible to *think* about them. They do not fit the definitions we use in physics: "position," "frequency," "velocity" and so on. The very experiments which seem to bring them so close only add to my confusion. I think of the tracks in a cloud chamber, the counter's click-pause-click. A thing was *there* when the counter clicked or the track was made. How can that thing not be *somewhere* at all times?

NEWCOMB: Yet you agree that what makes an interference pattern cannot at the same time occupy an exact position in space. The electron is not a miniature version of matter as we know it. You could shoot grains of sand through slits for the rest of your life and they never would make a picture for you. It is logical to expect a thing to be somewhere at all times but the electron is not a thing and logical assumptions about it often mislead.

OLDFIELD: The atomic world may not be pictured and logic is

bound to lead one astray. That is bad enough. But also you say that the results of your experiments always are inaccurate and the inaccuracy may not be gauged. In other words, the information you obtain is indefinite. From this, two things follow: first, predictions of the future also will be indefinite. Second, we never will know when the maximum information about the electron has been obtained. To pin down a wave property, we must design one kind of experiment; to detect a particle property, we must set up a totally different experiment. The same thing is viewed at two different times from two entirely different angles. When the results of the two experiments are studied, how can we tell whether or not something has been left out? The results of these experiments are inconclusive. They do not add up; they do not fit together like pieces of a puzzle.

NEWCOMB: But they do! In one way, it is misleading to talk about particular experiments as we have done, because they do sound "inconclusive." Actually our knowledge comes from a wide variety of different experiments and a mathematical analysis of them. For instance, we know that in some experiments we may measure the electron's position with the maximum accuracy possible. We know that in other experiments we may obtain an accurate measurement of its velocity. The different experiments are mutually exclusive: when one property is measured with precision, we cannot know the other at all. The interference experiment we talked about illustrates this general rule. Actually it was two mutually exclusive experiments, one done with position detectors, one done without them. When the recording instruments were used, we learned where the electron was; then we could not learn its velocity, for the interference pattern, which varies according to the electron's speed, gives us that information, and it did not figure in the experiment.

We can and we do design different sorts of experiments to catch different aspects of elementary particles; and the results of the different experiments add up. The puzzle pieces fit together. We know when we have obtained the maximum information possible.

At the same time, we are free to use our old and logical scientific definitions because we know when they do and do not apply. In physics, as you know, words refer back to measurements. Remembering the measurements which lie behind our words, we run no risk of contradicting ourselves.

OLDFIELD: I have been wondering why you speak to me of elementary "particles."

NEWCOMB: Yes. We use words like "particle," "wavelength" and so forth because we know when they are meaningful and when they are not. The same thing applies to models. No single model tells the whole story, but since we know what is left out we may continue to use them and as before they help us to understand. Going back and forth from one "contradictory" model to the other we begin to glimpse the whole.

OLDFIELD: But what about the other problem? You obtain the maximum possible information, but this still is not the information which is required for an exact prediction. To measure "position" one kind of experiment is necessary; to measure "velocity," another kind. Each measurement is accurate but each refers to a different instant of time. Only if they refer to the same instant is it possible to make an exact prediction about where something will be in the future.

NEWCOMB: Again the electron is not some "thing." The measurements you speak of apply to large-scale bodies and it is hardly surprising that they do not apply here. Neither do the comparable measurements which we must have in order to predict the future behavior of a wave.

OLDFIELD: Then on the atomic scale the door is closed to accurate prediction.

NEWCOMB: The door is closed to the kind of prediction which

can be made for large-scale phenomena; but that does not mean that our predictions are inaccurate. Electrons obey group laws. One could see that in the interference experiment. The wave pattern, of varying intensity, was composed of separate impact points. Where the pattern was most intense the number of impacts was greatest; where dimmer, there were fewer impacts; where dark, none. To make an accurate prediction where electrons are concerned one must consider a large group of them. As the number which hits the screen increases, the answer to the question, "Where will most of the electrons be found?" becomes increasingly accurate.

OLDFIELD: Your prediction strikes me as worthless. You knew when you set up the experiment what the result would be: a particular kind of interference pattern, depending upon the speed of your electron beam and the size and position of your slits.

NEWCOMB: I did not mean that we spend our time predicting what we know already! That was an illustration only of the kind of prediction we make with the motion laws which govern the atom, quantum mechanics. It is impossible to say where any particular electron will be in the future. But we are able to foretell, and very accurately, where a large group of electrons will be found. Our predictions always are statistical, their precision depending on how many similar cases are being considered. But this is not like going from door to door to find out how people are going to vote in an election, or why they buy certain products. Let me point out that the atomic region consists of similar cases. Electrons are abundant, you will agree! And all are exactly alike.

OLDFIELD: How do you know that they are all identical when you cannot measure the electron precisely?

NEWCOMB: The same way we know many other things in quantum physics: the mathematics says so and the mathematics works. It predicts what turns out to be so.

OLDFIELD: You were saying that the accuracy of quantum mechanics increases with the number of cases considered. That means quantum and classical mechanics must give the same solution for a large-scale problem involving many atoms.

NEWCOMB: The same solution, yes. But in a different form. The logical construction of quantum mechanics is such that only varying degrees of probability may be deduced from it. In the case of a large-scale problem, the probability given would be overwhelming—that is, it would amount to certainty. Quantum mechanics, then, is the general theory, classical mechanics being contained within it as a special case. That is why before there was such a thing as quantum mechanics Bohr's correspondence technique could be a reliable guide. With correspondence it was possible to obtain results which agreed with fact, but not invariably and not as a consequence of logical assumptions. Quantum mechanics, on the other hand, is a precision tool. Once we had that it became possible to calculate almost every atomic phenomenon. Today our understanding of how the atom behaves as a whole is very good indeed. We have gone on—to the fine structure of the nucleus which is manifested only under conditions of extremely high energy, when matter loses its atomic character.

OLDFIELD: I have been wondering how Rutherford was able to discover the nucleus. There was no correspondence then and no quantum mechanics. To explain his scattering experiments, he made some statistical assumptions, but he worked with the classical motion laws. He even was able to calculate the size of the nucleus.

NEWCOMB: Rutherford was lucky. In the particular case he was considering, the solution given by the old mechanics agrees with that given by the new. That is, the problem could be solved either way. Because of this happy coincidence, physicists began to follow the right scent in 1911.

OLDFIELD: What about Rutherford and Soddy's laws of radioactivity? They gave the rates of decay correctly. Therefore it must be possible to deduce the same laws from quantum mechanics if, as you say, that is the general theory.

NEWCOMB: Yes. The laws follow. And now we understand why they had to be statistical. We used to think that improvements in technology would make it possible to measure each individual atom in the way large-scale phenomena are measured, and then we would learn the answer to the question, "What triggers the nuclear explosion of radioactive decay?" Knowing that, we could work out the same kind of laws for the decay which we have in large-scale physics. We would get rid of the half-lives, the statistical rules behind them. Now we know better. The question, "What determines, or triggers, the decay?" is one which stems from knowledge of large-scale nature and has no meaning in the wave-particle world of atoms. We do not expect an answer to it; it was simply the wrong question to ask. The statistical rules of Soddy and Rutherford are not temporary. They express our knowledge of radioactive decay to the fullest possible extent.

OLDFIELD: Again, there are philosophical implications in what you say. It contradicts the idea that every event in nature in principle may be traced to a preceding cause. But rather than pursue this, let's talk about Niels Bohr's model of the atom. You have told me enough so that I think I can solve some of its problems for myself. Bohr's assumption that the electron is an ordinary particle and therefore would move in orbits about the nucleus was not correct; his idea of applying the quantum constant to the case of the atom was. By using the number h to limit the motion of a particle he could come closer to the mark than could classical mechanics, and then he could use correspondence to adjust his deductions. But to account for the motion of something with wave properties there would have to be new motion laws altogether.

NEWCOMB: Yes. And the same is true in the case of light. Quantum mechanics accounts for the wave-particle nature of matter; quantum electrodynamics is a comparable theory of light.

OLDFIELD: Going back to Bohr's atomic model, it strikes me that if you substitute a wave electron for the particle electron, you get a reasonable explanation for the way atoms behave; you get rid of the astounding properties which the particle electron had to have. From what you have said I understand that I am free to use a wave model so long as I do not assume that this model tells the *whole* story.

NEWCOMB: Go right ahead.

OLDFIELD: Well, take the simple case of the hydrogen atom in its normal state. The charge on the nucleus would bind a wave just as it would bind a particle. But a particle under those conditions could move in many different ways, many different orbits; a wave, however, could have but one motion, one shape. It would be confined within the area we call "the atom" by the nuclear field and it would fill this area. The wave must have a shape which fits. If it does *not* have this particular shape, this unique motion, then its vibrations will interfere with each other. The wave will be canceled out. It's the same sort of situation you have when a rope is held at either end and you shake it to set up a steady vibration. You see loops in the rope, their size depending on the rate at which you shake. But because of the boundary conditions you have set up, by holding the rope firm at both ends, you can only get 1, 2, 3 or some other whole number of loops which fit the length of the rope; there can't be any partial loops when the rope is in steady vibration. The boundary conditions restrict the shapes which are possible—here as in the case of the atom. Right away one begins to see a logical explanation for the Franck-Hertz type of experiment, which added more and more energy to the atom—that is, shook the rope faster. It is found that the atom would accept only certain definite amounts of energy. On the particle model that result meant that the electron had to find its way somehow to the proper orbit. But the wave electron by definition may accept only certain specific energies; otherwise it would cancel out.

Then there was the question about the shift in vibration frequency. The particle electron, falling back to the nucleus had to "know" where it was going and change its frequency of vibration to accord with its goal. That question does not come up with the wave electron. There are only the few possible wave shapes which will fit within the boundaries of the atom, only the few possible vibration frequencies which pertain to each shape. Automatically when the wave changes to another shape, it must change to the frequency which goes with that shape. It does not have to "know" anything. This wave model appears to answer the very questions which Bohr's model could not.

NEWCOMB: Yes. But let me remind you that the model does have definite limitations. When matter is subjected to exceptionally high energies, as it is in the sun, then the wave model applies no longer. Under those very high-energy conditions, the stability which characterizes the atom disappears altogether. The unique wave patterns which refer to the quantum states of matter no longer occur. But for the energies of our planet we need the wave model in order to understand; it accounts extremely well for the behavior of matter under "ordinary" conditions.

OLDFIELD: And it is logical! Nucleus of a specific charge means electron motion of a specific kind. No other is possible. If we go from hydrogen with one electron to helium with two, the conditions alter. The charge on the helium nucleus is larger and also the second electron wave would be affected by the first. This means that every helium atom in its normal state can have but one specific form, a form which is unlike that of hydrogen or of any other element. All electrons are identical, yet when one is added to the atom it means a radical change in properties. A change in quantity means a change in quality. Why? The particle electron does not tell us, or rather it tells us only if we assume that particles can exchange information across space as to their different motions so that none will duplicate that of the other. But the confined waves have to spread through the entire area of the atom, for that is the nature of a wave. They have to be everywhere and respond to conditions everywhere. Add an electron to the atom and a qualitative change follows naturally from this addition. I begin to see what you meant about the limitations of quantum physics enabling one to be logical!

NEWCOMB: Yes, quantum physics enables one to answer questions which begin with "why." Classical mechanics could not answer such questions, for its laws lead to an infinite number of possible solutions, not to the specific ones which account for what we actually find in nature. Why are the atoms in a metallic crystal arranged with geometrical regularity, as if they had been stamped out by a machine? Why, indeed, do snowflakes look the way they do? For that matter, why are all living things symmetrical? The answers lie in the quantum region of nature where Planck's constant is a significant number and where the strange union of properties which it announces is most apparent. The few electron wave shapes which may occur—the figure eight, the crosslike shape and so forth, all of them exactly symmetrical—lie behind and even are reflected in the forms we see. These few wave shapes, representing the motion which electrons assume in the nuclear field, explain why atoms and molecules combine in some ways and not in others. Thus the electron shell, due to Niels Bohr, is understood today as a pattern of interlacing electron waves. Some patterns, depending upon the number of electrons which are present in the atom, have missing parts, like missing pieces in a puzzle. An atom having such a pattern combines readily with atoms which can supply the necessary missing parts to form a tight wave pattern—and thus a molecule. When we come to the giant molecules of living matter, the structural arrangements are complicated and there are great numbers of them; but still the possibilities are limited. Only certain specific arrangements occur and in the DNA molecule that arrangement determines our specific genetic inheritance. This is a long way from the confined electron wave; but because we understand that so well, we may begin to understand the problem of life.

OLDFIELD: I would like to hear some more from you about quantum mechanics and have a look at the mathematics. But it is getting late and . . .

NEWCOMB: Before you go, let me describe the mathematical procedure we follow when solving a problem. I can sum things up by telling you what we actually *do* in atomic physics.

There are several symbolic forms of quantum mechanics, different translations of it, you might say. One of these bears a resemblance to the classical mechanics of moving bodies. Another form uses a differential equation similar to those worked out in the past to describe vibratory phenomena. But the symbols of quantum mechanics do not refer either to actual particles or to actual waves.

Most often, physicists use the wave form. Fundamentally it is a simple affair: there is the basic differential equation, the solution to which is a quantity which we call a "wave function." Just as in the past, one puts numbers into the equation, numbers which represent information gained from experiment and refer to such classifications as "position" and "frequency." We know that these classifications have a limited meaning in the quantum region. We know that our experimental information combines the effect of the measurement together with information about what is measured. The form of the equation is such that this limited meaning automatically is expressed. You will see what I mean more clearly when I describe the next step in solving a problem.

Very well. Numbers have been put into the equation, numbers representing an initial knowledge, and one wants to obtain an indication of what to expect in the future so that a prediction can be made and tested. Again as in the past one applies physical laws, solves the equation and obtains the solution, a wave function. This wave function does not describe real properties but the likelihood of finding these properties under particular experimental conditions. It represents an assortment of different events, some being more probable than others. The degree of probability will depend on the number of similar cases which are being considered.

OLDFIELD: Thank you, Newcomb. I will say "good-by" now. Frankly, this juggling of opposites we have been doing has made me feel dizzy and I would like to go somewhere and lie down.

NEWCOMB: It is often said of quantum physics that if you are *not* dizzy, you do not understand it. But while you are still upright, let me tell you about something which happened after quantum mechanics was put together. I will be brief. The fact that an electron's path cannot be followed suggests that it is not possible to distinguish one electron from another. Our equations, then, should express this indistinguishability; they should be constructed in such a way that electrons can be interchangeable without affecting the equation's absolute value. Technically, this may be done by making the equation symmetrical. Imposing symmetry on the equation, however, means limiting the solutions it can give. Then the question is: "Do the solutions given by the new equation agree with our knowledge?" and the answer is definitely "Yes" because the equation in symmetrical form limits the energy states which an electron may occupy, stating that no two electrons in the same system can have the same motion.

OLDFIELD: But that is Pauli's exclusion principle!

NEWCOMB: So it is. Before there was such a thing as quantum mechanics we knew that the principle had to be right. Now we see that it follows logically from the mathematics which represents our understanding of the physical world.

In the chapter which follows, we return to Werner Heisenberg, who by deciphering the code of spectra laid the groundwork for quantum mechanics, and to some of the others who contributed to this theory. We will be concerned primarily with the development of mathematical systems, systems which were worked out before their deeper significance was understood.

The development of quantum theory differed from that of relativity, for Einstein began with a general principle related to physical events and from this deduced logical consequences. What his theory was saying about nature was quite clear to him from the beginning. In the case of quantum theory, however, the underlying principle had to be dug out from the mathematical symbolism by applying this symbolism to many concrete cases. In the chapter after the next, we will come to this and to the real physicists who, like Oldfield, had to struggle with notions based on large-scale phenomena, notions sometimes called "common sense."

Back now to Heisenberg and to the time when no one understood what Newcomb explained in this chapter, when there was no quantum mechanics, when the experiments demonstrating the wave properties of matter had not been done, when the meaning of Planck's constant was unclear, and no one understood why it was necessary to rely upon statistical rules.

One thing stood out: the failure of logical reasoning. Ideas which seemed to fit everything else in nature failed to apply in the world of atoms and photons. But why this should be so was far from clear in 1924. There was always the question, "Does nature really not conform to the rules or in misapplying the rules is the failure mine?" Some physicists asked themselves a question which was even more disturbing: "Is it possible," they wondered, "that we have come to a part of nature which always will confound man's ability to understand? Is it perhaps something of which man's brain just cannot conceive?" Presumably Wolfgang Pauli was feeling something like this when he told a friend that physics is "really much too hard for me and I wish I were a comedian in the movies or something else like that and wish that I had never heard anything about physics."

This was just before "the dawn" when, suddenly, physics became easy again and one could say: It's so simple. Why didn't I think of it?

CHAPTER TEN

Creation of Quantum Mechanics

There is hardly any period in the history of science in which so much has been clarified by so few in so short a period.

—Victor F. Weisskopf

B Y MEANS of a lucky guess" it might be possible to transform Bohr's correspondence technique into a precise system. So speculated Werner Heisenberg in a letter to Wolfgang Pauli during the year 1924. The athletic young German who was suspicious of atomic models at an early age had been working at Bohr's Copenhagen institute with its first member who was also Bohr's assistant, Hans Kramers. From Kramers Heisenberg had learned how to use the correspondence technique. As we have said, this was closer to an art than a science; it could not be expressed in precise symbolic form but had to be learned by seeing how Bohr and his co-workers applied it in various cases. With characteristic boldness Heisenberg proposed to do something about this situation. He thought it would be possible to transform the art of correspondence into a logical chain of reasoning which would lead in every case to precise experimental findings. Such a logical system, such a precision tool, necessarily would describe the hidden world of the atom.

Heisenberg hoped to find this tool "by means of a lucky guess." He would not begin by imagining an atomic structure which could produce the phenomena which he wanted to explain and then work out the implications of this structure logically to see if they did in fact lead to the phenomena in question. Heisenberg's knowledge of atomic physics had greatly increased since he was a schoolboy in the Maximilian Gymnasium but he had learned nothing which strengthened his faith in models of the atom. He was keenly aware of the weaknesses in Bohr's atomic theory. That the discontinuous atomic energies postulated by Bohr did in fact exist experiment had verified, but there was no evidence at all for Bohr's quantum orbits. There was no evidence for the idea that the electron moved within the atom in the same way that larger material bodies move. By means of counters and cloud chambers the effects of the electron could be observed, but these electrons were free of the atom.

In fashioning his precision tool Heisenberg would avoid making any assumptions about atomic structure. He would work only with information that was closely tied to experiment, "hard" data. Roughly speaking, his reasoning went like this: ° We know nothing for sure about the electron inside the atom but we have a great deal of evidence for its behavior outside. We know that an accelerated charge (the electron) always produces electromagnetic radiation and that the frequency of this radiation always is the same as the "oftenness" with which the motion is repeated. (For example, the frequency of the electron's up-and-down motion in the radio antenna is the same as the frequency of the radiation which is sent out.) Let us call this hard information Proposition A.

Now the electron moving within the atom is an accelerated charge. According to Proposition A, we expect it to produce radiation and we expect the frequency of this to be the same as the frequency with which the motion is repeated. But when we calculate this on the assumption that the electron is moving in an

---

° It would be more accurate to say that "his reasoning could have gone like this." We do not pretend to do more than indicate the trend, or spirit, of Heisenberg's approach.

orbit the answer does not come out right. Orbital frequency is not equal to radiation frequency. Faced with this and other contradictions, Bohr invented the idea of jumps between orbits. He said that the energy lost in a *jump* determined the frequency of the emitted radiation. In the simplest case of the hydrogen atom this worked out. Nevertheless the idea of jump-energies rests on the assumption that there are orbits. We have no evidence that there are orbits but we have overwhelming evidence for Proposition A.

If we want to base a theory on that evidence rather than on supposition, there is a way out. There is a case where the predictions of Bohr's theory come close to Proposition A. That is when the electron is extremely far from the nucleus and, according to Bohr, its orbit would be enormous. There the jump-energies of his theory move toward a vanishing point. And when one calculates the frequency with which the electron would complete such a large hypothetical orbit, the answer comes out right: orbital frequency is equal to radiation frequency. In this borderline case between classical and quantum ideas which Bohr found, in this case where the electron is almost free of the atom and must move very much as it does when it is open to observation in the cloud chamber, we have firm ground on which to build.

Then suppose we do a detailed analysis of that case. We can make a breakdown of the kind of light which would arise at the atomic borderline and list so much light of the pure frequency X, so much of the pure frequency Y and so forth. This breakdown ought to be an exact symbolic representation of the electron's *motion* at the borderline of the atom, because the motion and radiation frequencies are the same. Now we want to find out about the motion in the atom's interior. There ought to be some logical way to deduce from this breakdown others which would represent that motion, because Bohr has shown us with correspondence that it is possible to reason your way from outside to inside the atom. It only remains to find the proper logical method and that should not be too hard because we already know the solution to which this method, if correct, will lead us. According to Proposition A, which is our "hard" evidence of the electron, our method must lead us straight to a frequency breakdown of the atom's actual line spectrum.

If we can do this, if we can proceed logically from the borderline case to the entire spectrum as it is observed, then we will have succeeded in breaking the spectral code. We will have found a system which exactly represents the unknown atomic interior. The problem is, how to do it. Perhaps (here was the lucky guess) there may be a clue in the motion laws which apply to large-scale bodies. In order to plot the path of such a body, one multiplies the quantity which designates the body's position at some instant of time by the quantity representing its momentum (mass × velocity) at that instant. Perhaps if we try multiplying one frequency breakdown by another . . .

Heisenberg was not in Copenhagen when he reached this point in his work which involved a curious rule of multiplication he had invented in order to manipulate the assortment of symbols pertaining to frequencies he had introduced. After a year with Bohr he had become assistant to Professor Max Born at the University of Göttingen in Germany; and it was there that he made the lucky guess. To work out his idea in detail he needed time. The opportunity came unexpectedly: in the spring of 1925 he had an attack of hay fever and was forced to take a vacation. Alone on an island in the North Sea called Heligoland, he thought hard, calculated incessantly and for relaxation climbed Heligoland's high red cliffs. In July he returned to Göttingen and handed his completed work to Max Born, who as his superior had to pass on it before it could be published. At the same time he asked for a leave of absence and soon he was back in Copenhagen working with Niels Bohr.

Heisenberg had not been able to deduce the line spectrum of an element from his new system. It was an enormous task of calculation under the best of circumstances (and still is, even with the help of a computer). The frequency breakdowns he had introduced added formidable complexity to what already was a very complicated problem. There was, however, some other evidence to support his work. Born recognized its importance at once and sent it off for publication.

But something about Heisenberg's system for multiplying one assortment of frequencies by another bothered him. It looked familiar; where had he come upon it before? His assistant was, Born said later, "very brilliant, but very young, not very learned." Had he stumbled on something that already was known? In Born's past lay the answer, as we shall see.

At the time when Max Born was racking his brains over Heisenberg's work he was the head of one of Göttingen's three physics departments. (German custom decreed that a professor must have a department, and since there were three physics professors at Göttingen it followed that there were three physics departments.) His theoretical work was widely known, as were the textbooks he had written, and young physicists flocked to Göttingen as they did to Bohr's institute. Pauli had studied there for a while; so, later, did such prominent physicists as George Gamow, Edward Teller and Robert Oppenheimer, as well as numerous well-known mathematicians, for Göttingen then was considered, academically speaking, the mathematical center of the world. It was known, too, for its surrounding countryside: the ruined castle where in an inner court coffee and whipped cream were served; the woods where an orchestra played and where, on a platform, students danced with the local girls under the big trees.

Sometimes, however, physicists criticized Göttingen. There was too much emphasis on purely technical mastery of physics, some people said—particularly those who were close to Niels Bohr. Also, due to the formalities of German academic life, Max Born was not as available for private discussions of physics as some of his students would have liked.

Born has been described as a moody, impulsive person, a gifted musician and writer (but one who found theoretical physics "more exciting than anything else"). He was a man known to speak harshly at times when his ideas were challenged, but a man who later would tell one, ". . . you must not mind my being rude . . . I have a resistance against accepting something from outside. I get angry and swear but always accept after a time if it is right."

Why was he puzzled by the work Heisenberg handed him in the spring of 1925? Because twenty years earlier, when he was a student at the University of Breslau, he had attended lectures in many different subjects, straying far beyond the academic confines of the physics student to visit classes in chemistry, zoology, philosophy, logic, astronomy, mathematics. One lecture had concerned matrix algebra; Born had not paid it much attention at the time, but years later a faint echo of the dimly heard lecture remained, and after hard concentration he was able to recall a few theorems governing matrices, square tables of numbers. Then he understood what had been bothering him about Heisenberg's new multiplication system: it was possible to formulate that system differently so that Heisenberg's frequency breakdowns appeared in a familiar square form—the form of the matrix. Unknowingly, the brilliant but not very learned young Heisenberg had rediscovered some rules of matrix algebra, the mathematical tool he needed in order to manipulate his assortments of atomic data.

That a form of mathematics invented long before could fit the newly discovered world of the atom was an amazing fact. Born, once he recognized it, put it to use. First he studied matrix algebra. Then, assisted by one of his best students, Pascual Jordan, he transformed Heisenberg's system into a broader theory. Later Heisenberg joined this collaboration.

In form, the extended system they worked out resembled classical mechanics, the laws governing motion of bodies, whether planets or the particles of matter on man's size scale. But while the symbolic vocabulary of classical mechanics refers to "position," "momentum" and other motion-defining quantities, the symbols of the new mechanics referred to breakdowns of atomic data, conforming to Heisenberg's original idea. By means of matrix algebra these tables could be manipulated in various ways to solve any problem which classical mechanics could solve, as well as the atomic problems which it could not. At least this was what the new matrix mechanics had been designed to do. Whether it actually—and invariably—led to fact remained to be seen.

When Heisenberg, in Copenhagen, saw some of the first results of Born and Jordan's work, he was bewildered. The learned men in Göttingen talk so much about matrices, he told Niels Bohr, ". . . but I do not even know what a matrix is." He managed to master the new system, however, and found that deductions from it agreed with known physical laws, such as that of energy conservation. "It was a strange experience," Heisenberg has said, "to find that many of the old results in Newtonian mechanics . . . could be derived also in the new scheme." But did it go further than the old mechanics? At first Heisenberg was disappointed; all his attempts to calculate the hydrogen spectrum on the basis of the new system had ended in failure. Then, to his surprise and delight the job was performed by Wolfgang Pauli, who had mastered the complexities of the recently published mechanics in record time. Not only was Pauli able to deduce the spectrum of the free hydrogen atom, but also changes in that spectrum produced by electric and magnetic fields, something which until then had not been possible. In due course spectra of other atoms also were derived and Heisenberg himself demonstrated the understanding of matter which matrix mechanics implied when with its help he predicted the existence of a form of hydrogen which was then "discovered."

Werner Heisenberg had succeeded in turning the art of correspondence into a precise logical system. He had broken the spectral code. And just as the particular kinds of light which atoms emit could be taken as a code to their internal behavior, so matrix mechanics which solved that code must speak in concise form of the atom. Bohr, Pauli, Heisenberg, Born and a number of others explored matrix mechanics in an attempt to learn its meaning.

Before they found it there was a new and unexpected development: in the spring of 1926 they were astonished to find, in the pages of a scientific journal, *another* new atomic mechanics. Like the one created by Heisenberg and his collaborators, it was a logically unified system, containing the results of classical mechanics within it and going further to account for the spectra of atoms. But this mechanics, proposed by an Austrian, Erwin Schrödinger, did not employ matrix algebra. Schrödinger had used an entirely different mathematical form, a form which had been developed in the past to describe vibratory phenomena, waves.

Now there were two ° entirely different systems (or so it seemed): one a wave mechanics, one a particle mechanics. Each was logically consistent; each led to results which agreed with observation. Physicists were baffled: which one described the

---

° Strictly speaking, there were three. As explained later in the chapter, someone else had developed a particle form of mechanics similar to that of Heisenberg and his collaborators.

Left: *Max Born in 1962, long after completing the work on quantum theory for which he won the Nobel prize; but not very long after the prize. That was awarded him in 1954.*

Right: *Louis Victor de Broglie, winner of the 1929 Nobel prize in physics for his discovery of the wave nature of the electron.*

reality? Ultimately they learned that both did and that the language of one could be translated into that of the other. The two were different expressions of the same thing. To this we shall return after a brief summary of the events which led to Schrödinger's formulation of wave mechanics.

In one way the development of this system paralleled the development of matrix mechanics: first (from the physicist-prince, Louis de Broglie, who played a role resembling Heisenberg's) came a penetrating idea. Then de Broglie's idea was taken up by Schrödinger, who extended and refined it to form a much broader theory, as Jordan and Born extended Heisenberg's idea.

But there was also a striking contrast in the way the two forms of mechanics evolved. Heisenberg's approach to the problem of deciphering spectra had been empirical, bold. Rejecting models, rejecting all assumptions which could not be tied to "hard" data, he had struck out into new territory, willing to go where the facts led him. His approach may be summarized by the words of a certain engine inventor: "Start her up and see why she don't run."

Schrödinger and de Broglie, on the other hand, did not intentionally discard a picture of the atom. Like Max Planck, they thought that it would be possible to account for the atomic realm within the general structure of classical ideas.

De Broglie has said that when he first proposed the idea which led to wave mechanics, he did not realize what its full consequences would be. Describing himself, he writes:

> . . . he who puts forward the fundamental ideas of a new doctrine often fails to realize at the outset all the consequences; guided by his personal intuitions, constrained by the internal force of mathematical analogies, he is carried away, almost in spite of himself, into a path of whose final destination he himself is ignorant.

He had taken this path when after serving in the First World War he returned to the study of physics.

As his title indicates, Louis Victor de Broglie was a member of a very old and noble French family. De Broglie was born in France, educated there (in that country it was not the custom for a student to train at other European universities) and after receiving his doctor's degree taught there. Even his military service was done in France, much of it, in fact, in the Eiffel Tower, where he served in a telegraph station.

It was in 1923 and he was thirty-one when he chose to develop for his Ph.D. thesis the idea which was his one major contribution to physics. During that year de Broglie had read about an experiment done by the American Arthur Holly Compton. Studying short-wave light (X-rays), Compton had noted a small but serious discrepancy between theory and measurement. Wave theory said that when a beam of light was scattered by an obstacle the wavelength of the scattered radiation would be the same as that of the original beam. This was conceived as a process fundamentally no different from the jouncing of a car on a bumpy road. In such a case, the people inside the car would bounce in the same rhythm as does the vehicle. According to Compton's measurements, however, when X-rays are scattered by a light substance (he used paraffin) the wavelength, in some cases, increases.

It was impossible to understand these measurements on the basis of classical wave theory, but Compton found that if he applied Einstein's theory of light quanta, and treated the scattering of X-rays as a collision between photon and atom, he could account very well for his findings. (For this discovery of what is called the "Compton effect" of light he later won a Nobel prize.)

Some years before these experiments, R. A. Millikan had announced the results of his careful work on the photoelectric effect, all of it in accord with Einstein's predictions. Now with Compton's discovery of a new process, a new effect which also was explicable only on the basis of Einstein's photon hypothesis, that idea began to seem less reckless. Physicists began to believe in the properties of light which were logically contradictory to its known wavelike aspect. And Louis de Broglie, pondering Compton's findings, was the one who asked the right question: "Can it be that matter also has a double—a dual—character?"

If this were so, if the hydrogen atom were an electron wave confined within boundaries imposed by the nucleus, then the shapes that this wave might assume would be limited in number. The atom then could have only certain definite energies. Thus de Broglie saw the possibility of accounting for atomic behavior on the basis of what was known about waves, on the basis of a model. That he ran into difficulties when he tried to carry out this program in detail will hardly surprise the reader of this book in view of the limitations on models which have been discussed. Nor will the fact that Erwin Schrödinger, who used de Broglie's wave idea and succeeded where he had failed, did so at the expense of a picture of the atom.

Historically, Albert Einstein formed a link between the work of Schrödinger and that of de Broglie. The ideas of the Frenchman appeared in his doctoral thesis and Einstein was one of the few who chanced to see that thesis, where the wave-particle duality, which Einstein had introduced in the case of radiation, was extended to matter. De Broglie's ideas were very close to Einstein's own; Einstein's investigations, comparing the properties of a gas with those of radiation, had brought him quite close to making de Broglie's discovery. Very soon Einstein published a paper in which he called attention to de Broglie's work, re-stated it in a forceful way and offered new arguments in its support. Said de Broglie: "The scientific world of the time hung on every one of Einstein's words. . . . Without his paper my thesis might not have been appreciated until very much later."

From that paper of Einstein's, Schrödinger, who was then teaching at the University of Zurich, learned of de Broglie's work and then, by inventing a form of wave equation, Schrödinger created the general system of wave mechanics. In his first paper, which appeared early in 1926, Schrödinger showed that the spectrum of the hydrogen atom could be derived from his wave mechanics, and it was this paper that took physicists by surprise. What were the waves which accounted so successfully for atomic behavior? How could an electron be both wave and particle? To support the wave idea of matter there was not a single piece of experimental evidence. Einstein could cite one experiment when he speculated in print that light might possess granular properties; to support the notion that matter had wave properties, de Broglie and Schrödinger could cite no observations whatsoever. Not until this idea was fully developed in theory was it confirmed by experiment.

The first evidence was found, curiously enough, by an experimenter who was not looking for it. He was an American, Clinton Davisson. Not having heard of de Broglie's idea, Davisson at first was unable to explain the results of some experiments which he had performed. At the same time Davisson was working on this in New York, another experimenter in Aberdeen, Scotland, was investigating the same problem. Neither man knew of the other's work. The experimenter working in Scotland was George Thomson, son of Rutherford's friend J.J.; and after learning of de Broglie's hypothesis Thomson set up an experiment to test it. He knew precisely what he was looking for: according to de Broglie, the electron's wavelength was proportional to its mass and its velocity of motion. The equation de Broglie had found said that:

$$\text{wavelength} = \frac{\text{Planck's constant}}{\text{momentum (mass} \times \text{velocity)}}$$

In order to test this hypothesis it was necessary to send a beam of electrons through something which would bend (diffract) it to produce an interference pattern—if the suspected wave properties did in fact exist. De Broglie's equation said that the wavelength associated with the electron, even when traveling at a relatively low speed, would be extremely short, on the order of atomic dimensions. The diffracting device, therefore, would have to be atomic in size. George Thomson used thin films of metal in crystalline form. Because the atoms in a crystal are lined up in parallel rows, they can function as a kind of diffraction grating, one which would be exceptionally difficult to create artificially.

When Thomson sent electrons moving at a controlled speed through a crystal grating to a photographic plate on the other side of it, he obtained an orderly pattern which looked almost exactly like the one made by short-wave light under similar conditions. This pattern told him the electron's wavelength; it was precisely the size which de Broglie had predicted.

The father of George Thomson had been first to measure the electron's mass and charge; now the son measured its wavelength. But he was not the first. Davisson had obtained similar results a few months earlier. He was working then in the laboratories of what now is Bell Telephone Company, experimenting there with beams of slow-moving electrons to learn how they are scattered when striking the surface of different metals. It was a problem in pure science or what today would be called "basic research," and Davisson had chosen it for himself. He was an exceptionally gifted experimenter and the firm allowed him considerable freedom, in a day when industry rarely subsidized basic research.

Davisson's research had been going on for several years when in 1925 a fortunate accident took place. He was investigating the scattering pattern produced by a beam of electrons sent at a sheet of nickel when accidentally the surface of the nickel specimen became oxidized. To restore the surface the specimen had to be heated, and in the process it became so hot that the surface atoms were changed to form a few large crystals. Therefore when the experiment was resumed and another electron beam was sent at the nickel target it was in effect being sent at a crude crystal grating. In a way, Davisson was performing the experiment which Thomson, later and independently, would perform on purpose.

Because he was working under exceptionally good conditions, Davisson obtained a clear-cut experimental result, a result which must have struck him as fantastic. Davisson had not heard of de Broglie's hypothesis; as far as he knew, he was experimenting with a shower of particles and having sent these particles at an obstacle, they had been scattered—not every which way, as one would expect, but only in a few well-defined directions. Davisson did "not fail to keep his eyes open for the possibility that an irritating failure . . . to give consistent results may once or twice in a lifetime conceal an important discovery." He took the experimental result seriously, and with one theory and then another tried to explain what had happened. At length, after consulting physicists in Europe, he was led to de Broglie's hypothesis. Then, with an improved technique and the assistance of Lester Germer, he performed additional experiments and obtained results which were in excellent accord with de Broglie's hypothesis.

Recently a physicist has said that even today it would be very difficult to duplicate Davisson's experiments. They were, he said, "a triumph of experimental skill." The physicist who speaks here is George Thomson. He came to know Davisson as well as his work after 1927, when the results of that work were published. Davisson was a man, said Thomson, of "exceptional attraction . . . an individualist, working with at most one or two other people and doing much with his own hands." He was, Thomson adds, "full of quiet fun, sometimes unexpected."

The experiments of George Thomson and Clinton Davisson (for which they shared the 1937 Nobel prize in physics) demonstrated the wave property of electrons; later experiments would reveal that protons, atoms—even, under certain conditions, molecules—behave as if they are what de Broglie termed "matter waves" and what other physicists call "de Broglie waves." But, to repeat, all of these experiments came later, after the theory which accounted for their results was mathematically complete, and even after the meaning of the mathematics was understood. Before going on to the discovery of that meaning we first will conclude our account of how the mathematical tools of quantum mechanics were developed; for the name Paul Dirac has not yet appeared in these pages, although his contributions were major.

Dirac's work was abstract, decidedly so; and in Dirac's opinion that was as it should be. The job of a theoretical physicist was done, he thought, when his abstract symbols led to fact. To seek the meaning of these symbols was in his opinion something else, philosophy, and it did not interest him. "Beware of forming models or mental pictures at all," he once admonished Erwin Schrödinger.

Dirac worked alone. Although he studied for a while at Bohr's institute, he participated but rarely in the eternal group discussions. Indeed, he spoke but seldom. In a deserted classroom, Dirac would be seen, doing nothing, just sitting. "One must think hard," he said once, when asked to explain his occupation. If the observer waited, he might see Dirac, at last, write something down. But by that time his work was done; he just was transcribing the result.

Physicists said that this usually was correct the first time and that it was also remarkably simple and original. They spoke with admiration of Dirac's style, meaning his line of reasoning.

While this was expressed in advanced mathematical language, it is not absolutely necessary to know the language in order to glimpse the style, for it showed itself outside as well as inside physics. It was evident in the way Dirac conversed. "One never should begin a sentence," he said, "until one knows what the end of it will be." On the quite rare occasions when he said something, he tried to express himself with the utmost succinctness. Upon being told, for example, that he might write the phrase "Not to be published in any form" on some scientific material which he did not want released to the press, Dirac appeared dissatisfied. When asked what was the matter with the suggestion, he explained that in the phrase he had been given the words "in any form" were not necessary.

Even a "yes" or a "no" could be too wordy, in Dirac's view.

ULLSTEIN

Paul Adrien Maurice Dirac at the age of thirty-one. A year earlier he had become Lucasian Professor of Mathematics at the University of Cambridge, the chair once held by Isaac Newton.

Once, when being served tea, he was asked whether he would take sugar. Having replied "Yes," he was surprised at the query which followed: "How many lumps?" It was unnecessary to ask that, he thought. Sugar came in pieces, each one of the same specific size. A piece of sugar, then, was the definition of how much sugar. When he had said "Yes," he would take sugar, it could mean nothing else but one piece.

The physicists who were drinking tea with Dirac on that occasion and who by artful questioning were able to learn his view of sugar were delighted. They recognized the same thinking style which they knew from his work in physics.

Like Schrödinger and like Heisenberg and his collaborators, Dirac also had independently of the others created a new atomic mechanics. He had done it in England (his native country) and he had done it at the age of twenty-three. His work sprang from Heisenberg's idea, just as did Born and Jordan's, but while they had seen the crucial point only after considerable preliminary work, Dirac had recognized it virtually at once. In Heisenberg's system frequency breakdowns were substituted for such mechanical quantities as position and momentum, quantities which are represented in classical mechanics by the letters q and p. And according to Heisenberg's system p multiplied by q was not the same thing as q multiplied by p. Unlike classical mechanics, unlike ordinary arithmetic, the order in which the two quantities were multiplied affected the product's value. Suppose, reasoned Dirac, that Heisenberg's rule expresses the one fundamental difference between the unknown laws which govern the atom and the classical laws. In that case if one knew the difference between $p \times q$ and $q \times p$ and if this difference always was the same, it would be a simple matter to transform any equation of classical mechanics into a comparable equation referring to the atom.

With this in mind, Dirac looked for some known mathematical technique which he could adapt to his purpose. He found what he wanted, the Poisson brackets, and applying that technique to Heisenberg's system, was able to calculate the exact difference between $p \times q$ and $q \times p$, finding that indeed this difference always was the same. His object was reached: using the Poisson brackets one could transform any classical equation into a comparable quantum equation. Thus a new mechanics which retained the structural coherence of the old was fashioned, by Dirac, in one stroke.

The difference between $q \times p$ and $p \times q$ referred to above was determined by Planck's constant. When Dirac's mechanics was used to solve a large-scale (or many-atom) problem, p and q would be large quantities. Comparatively, the value of h would be close to zero and $p \times q$ could be taken as equal to $q \times p$, just as in classical mechanics. In other words, Dirac's form of quantum mechanics, like the others, contained Newtonian mechanics as a special case.

Just before Dirac's ideas appeared in print, Born and Jordan (collaborating with Heisenberg by letter) had come to the same idea. Their first paper, published prior to Dirac's, had developed the use of matrix algebra; then, using not the Poisson brackets but another technique, they too had found the difference between $p \times q$ and $q \times p$, and knowing that, they were able to formulate their unified theory, matrix mechanics. Thus Born and his collaborators and Dirac had the same idea at almost the same time; and soon afterward came the third version, Schrödinger's wave mechanics, which he had developed independently of the other two.

A hectic period followed during which the originators of the three different versions—together with other theoretical physicists, Pauli prominent among them—perfected their own and each other's work, applied it to the solution of problems and demonstrated that the symbolic vocabulary of any one version could be translated into that of the others. Heisenberg's multiplication rule, for example, which was at the root of the two forms of particle mechanics, was also implicit in Schrödinger's wave mechanics. The three different but equivalent schemes came to be known by the general term "quantum mechanics." Of the three, physicists found Schrödinger's the most convenient to use. Dirac made changes in this: applying concepts of special relativity theory, he wrote Schrödinger's wave equation in a slightly different form. The new equation implied that the electron possessed what is called "spin," ° although Dirac had not designed his equation to say this. And there already was evidence that the electron possessed just such a spin. But the relativistic wave equation implied something else as well: that the electron (and other elementary particles, for it described them all) exist in pairs, each having a twin "antiparticle" with the same spin and mass, but opposite charge. Because of *this* impli-

° The word "spin" is meaningful under the experimental conditions when the electron may be pictured as a particle.

cation Dirac's formula was not at first taken seriously—until the positive electron (positron) was discovered, followed one by one by identifications of other antiparticles. As well as refining the mathematical theory of matter, Dirac introduced a quantum theory of radiation which became a partner to quantum mechanics.

This word-sketch of how the tools of quantum mechanics were shaped is complete now, and we turn to the exploration of its meaning, to the particular efforts of Schrödinger, Bohr and Heisenberg to interpret the mathematics and learn how it described the atom. The understanding reached has been presented already, in the dialogue between Newcomb and Oldfield. Now we will see how actual physicists came to it. Following this chapter, we return at last to Albert Einstein, to learn what happened after his decision to return to Germany and how he received the new interpretation of quantum mechanics.

CHAPTER ELEVEN

Interpretation of Quantum Mechanics

We agree much more than you think.

—Niels Bohr

I N COPENHAGEN, Niels Bohr studied the new publications of the Austrian physicist Erwin Schrödinger in which he presented his wave mechanics. Bohr was mystified. By means of a differential equation similar to those which were used to describe wave phenomena, Schrödinger had been able to decipher atomic spectra. He had been able to account for the atom's discontinuous energies without bringing in quantum rules to limit—and contradict —classical physics. What did the wave equation signify? Schrödinger appeared to think that it referred to actual wave phenomena and that all the contradictions of quantum theory could be resolved on this basis. Did he really believe that this was so?

That he did Bohr soon learned. The wiry, intense Austrian physicist long had sought just such a resolution. It was his great, his passionate wish to eliminate what he once called "gross dissonances in the symphony of classical physics" and all of these sour notes were due to the idea of energy quanta. Planck had used the idea to limit the infinite number of solutions given by classical physics and so explain the specific energy distribution of black-body radiation. Bohr had used the idea to single out specific energy states from the continuous range given by classical physics and so account for the atom. But the idea of energy portions, while it answered some questions, always had raised others. What was in back of the energy bursts of black-body radiation? What triggered them? Why could the atom have some energies but not others? In every case where the quantum had entered physics, questions like these had followed, questions about the deeper cause which brought about discontinuity. Such questions appeared to be the price one had to pay in order to bring theory into line with fact.

But suppose the price paid was understanding itself? So Schrödinger felt. The laws of classical physics make it possible, at least in principle, to analyze any part of the physical world to answer the question, "How does it come about? How does it work?" Just as a watch may be opened up and a study made of the tiny inner parts which determine its operation, so the classical laws enable one to understand physical processes.

Even light and electricity, which cannot be described in mechanical terms (that is, in terms of a material carrier, the ether), nevertheless can be analyzed in order to understand how one electromagnetic wave follows inevitably from that of the preceding instant. Classical physics enables one to divide up the process and see its causal development.

It is true that before the classical laws are applied in order to learn the evolution and outcome of a particular event certain things must be known. For example, before one can predict the trajectory of a shell shot from a gun one must know the shell's speed as it leaves the gun and the position of the gun barrel. Once these so-called "initial conditions" are known, however, the classical laws make it possible to follow the evolution of the process step by step, cause followed by effect which in turn produces another effect, in sequence and consequence. Thus inanimate nature could be conceived as a giant watch, or as a motion picture which may be slowed down so that the "action" at any point is traceable to that of an earlier or a later point.

It was this symphony of classical physics which the energy discontinuities had interrupted. A process which changes by jumps may not be understood in a causal—or more strictly speaking, a deterministic—way. Energy quanta break the cause-and-effect chain. They make exact prediction of the future impossible in the world of atoms. It had been necessary to substitute the prediction of what is more, or less, likely.

Schrödinger's wave mechanics was the culmination of a long search to find some other way to solve the problems which already had yielded to the quantum idea and thus rid physics of the sour notes. Although quantum rules brought theory into line with fact, Schrödinger was not convinced that this was the only way to do it. Perhaps energy was not quantized at all but merely appeared so because viewed with ideas which were not appropriate. Perhaps it would be possible to deduce the same facts on the basis of different ideas, ideas which restored continuity to the atomic world so that it could be understood in the old clock-like, deterministic way. This was Schrödinger's objective. "As to atomic theory," he said, "I tested and rejected many an attempt (partly of my own, partly of others) to restore at least clarity of thought. . . . The first to bring a certain relief was de Broglie's idea of electron waves. . . ."

Schrödinger was almost forty years old when building on de Broglie's idea he found a way to deduce the hydrogen spectrum without quantum rules. His wave mechanics was—like the other versions of quantum mechanics—a closed system, meaning that *in principle* all physical happenings on the atomic scale could be interpreted on the basis of its abstract logic. How to apply this logic to concrete cases was yet to be established; physicists would occupy themselves with this problem in the years ahead. Also the path from mathematics to the prediction of fact was long and arduous, so much so that some problems could not be worked out at all; the amount of calculation involved was prohibitive. Still, what was evident in 1926 later was substantiated. Although Schrödinger's fundamental wave equation was revised by Dirac, its essential features remained and the solutions to which it led were found to agree with fact, either known or subsequently discovered.

What interested Schrödinger, however, was not so much the application of his work as the form it took. The fundamental tool of wave mechanics was his differential equation, which when applied to a problem gives the solution in terms of the symbols ψ (x, y, z). The Greek letter ψ (psi) may be interpreted as a wave disturbance defined by x, y, z, which refer to three dimensions of space. Equations of this sort are part of the equipment of classical physics. By their means it is possible to analyze continuous phenomena such as a train of waves, and understand how at any given instant the wave is determined by its state at an earlier instant.

Then Schrödinger, so it appeared, had reached his goal. Without quantum rules, without making energy discontinuous, he could account in principle for the whole range of atomic phenomena. The jumps which prevented a causal analysis had been eliminated. One might assume then, as did Schrödinger, that matter fundamentally is a wave phenomenon, and that what appear to be elementary particles are merely the tiny regions of space where the waves reinforce each other.

This was what mystified Niels Bohr. First of all there were the experiments which could be explained only on the basis of something discontinuous. It was impossible to understand how wave interference could cause the click of a counter or the scintillation on a screen. Apart from experiment, the symbolism of wave mechanics suggested strongly that the waves to which it referred were far from being actual wave vibrations. With Schrödinger's mechanics one could represent a single electron as a wave disturbance in a space of three dimensions, but only a single electron. Two electrons could be represented only in an imaginary mathematical space of six dimensions; three electrons required nine spatial dimensions, and so forth.

For reasons like these Bohr told his Copenhagen friends, "I do not understand the wave mechanics," and hoping that the inventor could explain it to him, he invited Schrödinger to Copenhagen. There the Austrian physicist spoke at seminars to the members of Bohr's institute, and afterward for the sake of the head of the institute continued his explanation of wave mechanics in a series of discussions which went on late into the night.

Instructing Bohr could be a long and even painful process. In the beginning the Danish physicist with the big head and heavy, drooping features would ask in his shy, quiet way, a simple, almost childlike question which seemed to stem from a vast ignorance of physics. Schrödinger, who moved, and also thought, very quickly would answer Bohr's question with clarity and even eloquence, for although, in contrast to Bohr, he preferred to work with mathematical symbols as much as possible, it was easy for him to put his thoughts into words. (He even could write poetry.)

After Schrödinger had answered Bohr's question, often speaking at considerable length, he would be greeted with another question, equally naive, for Bohr as usual was thinking out loud, a process which involved trial and error, or as one physicist has put it, "progress by making almost all possible mistakes, the great point being only to make them as quickly as possible and to learn from them."

This mode of thought was not necessarily foreign to Schrödinger but he did not share Bohr's habit of thinking something through in the company of another. He did not mix much with other physicists and had no collaborators. He was, however, like Bohr, a good athlete and the two men could maintain their intellectually demanding talk for long stretches of time. Gradually

Erwin Schrödinger at about the time of his visit to Copenhagen.

Bohr's study would darken with smoke from his pipe, which he always was struggling to keep lit. It was impossible to talk and at the same time relight the pipe. Again and again, after getting it started, he would not wait long enough before resuming his argument and the pipe would go out. Much of the time he was holding a lighted match, which gradually burned down toward his fingers (those who talked with him watched it with fascination), but was dropped just in time. Beside his chair a pile of matches would accumulate. Finally Bohr would get down on all fours and clean up the debris he had made.

As time went on, and with Schrödinger's help, Bohr began to understand wave mechanics, his questions became harder to answer. He was challenging Schrödinger on a fundamental issue, that closest to the Austrian's heart: Had his wave theory indeed restored continuity to physics? Was the idea of quanta indeed no longer necessary? This led to specific cases. What, for example, of the first use of the quantum? In order to write a formula which agreed with observation, Planck had had to assume that radiation was sent out from a glowing solid in bursts, or jumps.

Now, asked Bohr, could Schrödinger account for black-body radiation without bringing in the jumps? Schrödinger tried to do it. With the mathematical equipment of wave mechanics he could write a formula for black-body radiation which, like Planck's, agreed with observation. But this did not solve the problem: even in wave language the jumps remained. In the specific case Bohr had chosen, this fact was glaringly evident. It had not shown up in Schrödinger's wave interpretation of the atom; but there also, as a close analysis revealed, the same thing was true. The atom's transitions from one energy level to another were events which could not be analyzed further. Whether the electron was conceived as an orbiting particle or as a wave vibration the discontinuities remained. Schrödinger's theory had disguised the sour notes; it had not eliminated them.

Schrödinger, by this time, was exhausted. "Had I known," he said darkly, "that we were not going to get rid of the damned quantum jumping, I never would have involved myself in this business."

Bohr replied to the effect, "*You* may be sorry but we, who have been learning from your work, are happy that you did."

Bohr had found evidence that Schrödinger's interpretation of wave mechanics was not correct. The findings of Max Born settled the matter conclusively, for he was able to show that elementary particles could not be created by wave interference as Schrödinger had suggested. The psi wave, in its imaginary space, was not to be interpreted as the description of an actual wave vibration. It was not to be identified with the matter (or de Broglie) waves first revealed by the experiments of Davisson and Germer.

Although Schrödinger had failed to realize his desires he did not relinquish them altogether. It remained his conviction that some day the quantum idea would be eliminated from physics and determinism restored. Always he could reply to the points made by Bohr, Born and others, "I am unable to answer this, I grant, but I believe that eventually a way will be found to answer it," just as he could say, "It is true that our instruments seem to detect particles, but this may be due to some interaction between wave and instrument which is not yet understood."

Another great physicist, Max Planck, shared Schrödinger's view; and when the older man retired from his post at the University of Berlin he recommended Schrödinger as his successor. Not long after his visit to Copenhagen, Schrödinger moved to Berlin, where Planck in a speech saluted him as the man who had reestablished determinism in physics, ending the crisis brought on by his own quantum theory.

In Copenhagen, meanwhile, ideas were taking shape which would further undermine Schrödinger's. The long debate, which had left the Austrian tired and gloomy, had had quite the opposite effect on Niels Bohr, who thrived on the long arguments which sent others from Copenhagen in a state they described as "total exhaustion." At last, Bohr felt, he was beginning to understand the relationship between the abstract logic of the mathematics and real events.

Bohr was beginning to think that the atom and its components were not the particles considered by the old mechanics just as they were not the waves of the new wave mechanics. Neither the symbols p and q nor the symbols $\psi\,(x,\,y,\,z)$ should be understood as literal descriptive labels. Yet they should not be dismissed as meaningless abstractions either; the symbols that appeared in the new quantum mechanics and the logical relationships among them must refer in some way to actualities. Bohr thought that he knew the way to learn what the abstractions were saying. The mechanics of Dirac and of Heisenberg and his collaborators had been patterned after Newtonian mechanics, and in back of that mechanics was a particle picture. It must have a real, although limited, validity. The same could be said of the wave picture in back of Schrödinger's mathematical form. Certainly that picture provided insight into the behavior of electrons within the atom.

Full understanding would come, Bohr thought, from a comparison of the two pictures. Long ago he had been attracted by the contradictions implied by the nuclear atom and in his work deliberately had stressed those contradictions. Now he would do the same, comparing the logically correct particle form of quantum mechanics with the equally correct wave form of quantum mechanics which appeared to contradict it. He was optimistic. Schrödinger's view and his own might seem far apart, but that was only because part of the truth was hidden. "We agree much more than you think," he had told Schrödinger, at the same time that he had faulted his argument (a remark which probably did not comfort Schrödinger much). And after the Austrian's departure, Bohr became wholly absorbed in an effort to dig out from the symbolism the meaning which he felt was there but did not see clearly and could not demonstrate.

Absorbed in the same problem at the same time and in the same place was Werner Heisenberg, on leave of absence from Göttingen. He had participated actively in the discussions between Bohr and Schrödinger. With Bohr he had joined in arguing that wave mechanics had not gotten rid of the discontinuities which made exact prediction impossible, but he did not share Bohr's feeling that a wave picture of the atom contained elements of the truth. The experiments which demonstrated the existence of matter waves, thus vindicating Bohr's feeling, had not been done. Heisenberg felt, and felt strongly, that wave mechanics was a mathematical contrivance which contained no clues to real happenings. Like Bohr, he believed that elementary parti-

cles were abstract, defying classifications based on man's ordinary experience. To find out about these abstractions one
should avoid untested assumptions and stay very close to measurable phenomena. One should study the symbolic system which
had been constructed out of reliable data and clearly must in
some way describe the atom—his own matrix mechanics.

He tried to convince Bohr of this, but the Dane was equally
sure that his own plan of attack was the one which would lead to
understanding and argued its virtues at length to Heisenberg, refining his ideas in the process. (Bohr often was teased by his associates for this habit of nonstop talk, useful to him but trying
for someone who also had something to say. A cartoon drawn by
George Gamow shows Bohr with another physicist, also inclined
to be very talkative, who has been tied to his chair and gagged.
"Please, please," says Bohr to his victim. "May I get a word in?")
Heisenberg, however, like everyone else who knew Bohr, did not
hesitate to interrupt the older man, as loudly as necessary, and
staunchly continued to defend his ideas. The argument grew hot;
the views of the two, although different, were not *very* far apart
(not as far, for example, as Bohr's and Schrödinger's). It was an
argument between members of the same family and hot, perhaps, just for that reason.

Although neither man succeeded in changing the other's mind,
they were able, at the same time, to work together. Their chief
objective was the same: to force from mathematics an understanding of the atom. To do this, it was necessary to test one interpretation of the symbolism, and then another, by examining a
great number of different experiments. They had to learn, for example, how one could describe an electron moving through a
cloud chamber with the new forms of mechanics. By working out
such problems and seeing the form they took mathematically,
various interpretations were tested. Each could use this common
work to advance his own attack—attacks that differed in method

Niels Bohr on one of his long walks devoted to the discussion of physics. On this occasion in 1931 the walk is along the Appian Way outside Rome and his companion is a young Italian who liked to walk, too, Enrico Fermi.

as well as in conception. For Bohr proceeded slowly but very surely, going over and over the same ground, doubling back to make sure that nothing had escaped him and that he understood from many different points of view. Heisenberg, on the other hand, advanced by great leaps and pounces, which often—but not always—brought him closer to the goal.

Almost every day, during the fall of 1926 the two physicists arranged to meet and discuss experiments, discussions which went on and on. Meals were not an interruption; they talked physics during them (and when another physicist sat down at their table in the institute lunchroom, he would contribute to the subject at hand). Because a blackboard was necessary they worked indoors much of the time, but they did not like to remain stationary for long; the work would continue as they strolled in the park back of the institute or even as they rode horses together. Until the Tivoli amusement park closed for the winter they went there often, particularly to see at the ball-throwing stand who could outscore the other.

Left: *Werner Heisenberg, on the left, during a ski-trip in the Bavarian Alps in 1931, the year before he won the Nobel prize. With him is Felix Bloch, also a member of Bohr's institute and also winner of a Nobel prize.*

Right: *Another snapshot of Heisenberg in the Alps. The goggled individual reading on a rock is Niels Bohr.*

The collaboration which began in September continued until the following February. Progress was slow; one hypothetical interpretation had to be discarded and then another. When in effect they put a question to nature, the answer they received made no sense, for the question had been phrased in the wrong language. But they knew no other. As physicists, they possessed a large vocabulary of technical words with which to describe the processes of nature, but their words referred back, in the end, to a fundamental distinction. A physical process could be analyzed in terms of moving particles or in terms of the propagation of a wave; language referred either to one case or the other. What is more, language implied that it was possible always to distinguish between the two cases. In back of words was the logical assumption that given two statements:

This is a particle.
This is not a particle.

one of the two must be correct, just as given the two statements:

This is a table.
This is not a table.

one must be correct. That being the situation, it was close to impossible *not* to ask loaded questions, questions which assumed that even though the electron displayed the character of both a wave and a particle, it "really" was one and not the other. Before progress could be made, Heisenberg and Bohr had to become distrustful not only of words but of intuition—"common" sense. Heisenberg has described his feelings at this time:

> I remember discussions with Bohr which went through many hours till very late at night and ended almost in despair; and when at the end of the discussion I went alone for a walk in the neighboring park I repeated to myself again and again the question: Can nature possibly be as absurd as it seems to us in these atomic experiments?

Physics has been called "a game in which the scientist puts questions to nature, in hopes of receiving a reply." But nature, the physicist's opponent in the game, seldom is moved to speech and possesses infinite patience. When repeated efforts have failed, when despite the cleverest queries the opponent remains silent, the questioner begins to lose hope of ever winning this game. Then it is hard to go on. ("Of course I am unhappy," said Pauli. "I do not understand. . . .") At such times the optimism of a Rutherford or a Bohr is a distinct asset.

In the winter of 1927, however, even Bohr was not feeling cheerful. For months he had been worrying the same problem by day and dreaming it by night. He was worn out; it was hard to concentrate. Interruptions to his train of thought, which as a rule he did not mind, began to bother him. He felt, too, that the collaboration with Heisenberg, valuable as it had been, had gone on long enough. He was weary of arguing his differences with the younger man; he wanted peace.

In February he decided to take a vacation, go to Norway and ski. But in the mountains of Norway, as in Copenhagen, his thoughts remained focused on the same problem: utilizing the information gained from his common work with Heisenberg he continued his mental play with the contrasting wave and particle pictures of the atom. In a few weeks he returned to Copenhagen tanned, refreshed and much closer to the resolution which he—and Heisenberg—had sought so long.

Meanwhile, in the absence of Bohr, Heisenberg had been able to collect *his* thoughts, and pursuing *his* original plan and utilizing the results of the work with Bohr, he too had made great strides forward. He could greet Bohr with some ideas which he had taken care to submit first to Pauli's critical eye. And Pauli's opinion had been given: "Day dawns on the quantum theory."

Heisenberg had discovered the fundamental principle of quantum mechanics, the base from which all the rest logically follows. This principle (also called "a law") takes the form of a mathematical relationship between certain definitions that are used in physics. It is called the "principle of indeterminacy," or "uncertainty." Bohr's contribution was quite different. He was concerned above all with meaning and therefore with the use of language and logic. His work, in its final form—which came after he returned to Copenhagen and learned of Heisenberg's law—was an interpretation of that and of the entire mathematical expression of quantum mechanics.

Since Heisenberg's principle lies at the root of Bohr's interpretation, we first will speak of that. The turning point in Heisenberg's search had come when finally he asked the right question. Until then his thoughts had been focused on the problem: "How can various experimental situations be expressed by means of the matrix symbolism?" Suddenly it occurred to him to ask the question in reverse: "What if the only experimental situations which *can* occur are those which fit the logic of the symbolism?" His rule of multiplication said that the order in which p and q are multiplied determines the value of the product. Suppose that in the case where the symbols refer to position and momentum, one took "the order in which multiplied" to mean "the order in which these are defined in the experimental situation, the order in which they are measured." This would seem to suggest that when in a given experiment position is defined exactly, its companion, momentum, cannot be.

With this as a hypothesis, Heisenberg made a fresh study of experiments in which q and p are measured. He found that in case after case, experiment after experiment, the idea held up. And he was able to calculate from these experiments a certain minimum quantity which always was the same, a constant. This quantity represented an inevitable uncertainty in the measurements which define p and q; and the quantity in question was the one which first had been identified by Max Planck. Now, as we will see, it took on new meaning.

From the analysis of a wide variety of different experiments Heisenberg distilled his law which states that any measurement of position and momentum, any measurement of energy at a specified instant of time, must result in uncertainty equal *at least* to Planck's constant, or

$$\triangle q \times \triangle p \geqq 6.6 \times 10^{-27}$$

According to this law of indeterminacy, q and p are not independent of one another: if one is semidefined, so must the other be. If one is altogether unknown, then the other is defined exactly. They are, Heisenberg said, like "the man and woman in the weather house. If one comes out, the other goes in."

In the relationship which Heisenberg had found lies the fundamental difference between classical and quantum physics. The classical laws make it possible to predict where a moving body will be found in the future if—and only if—one is able to obtain certain initial information: the position and momentum of the body at an earlier instant of time (as in the illustration of a shell trajectory, given earlier). The comparable information which one must have in the case of a wave is its energy (which depends upon frequency) at a particular instant of time. Only if these precise quantities are known can one analyze the precise causal development of a wave vibration. Heisenberg's law denies that this initial information may be obtained, stating that q and p are inseparable to the minimum extent of the quantity h and that it is not possible to allow for this inseparable relationship, for although it never is less than h, it may be slightly more.

Classical physics is valid, then, only when the quantities to which q and p refer are very large relative to h and the connection h represents can be ignored. In the world of the atom, however, Planck's constant is a significant number, and the discontinuities it introduces make exact prediction in the classical sense impossible. The future still may be foretold, but not in the old way. Statistical reasoning, based on a great number of like cases, must be introduced.

This, as we have seen, was the path physics had been forced to take in order to account for the atom; now Heisenberg had found the reason why. Paul Dirac, Max Born and his collaborators had recognized Heisenberg's rule for multiplying p and q as the crux of his matrix system and this recognition had enabled them to construct the generalized system of quantum mechanics. What they had done for technical reasons could have been done earlier if there had been an indeterminacy principle.

But what was its meaning? Why did p and q escape simultaneous definition? What was the significance of the quantum constant first identified by Max Planck? And did the law mean a limitation to scientific knowledge? Had the scientist in his attempt to understand nature in greater detail come at last to a barrier? The law could be so interpreted for a limitation to causal analysis could mean a block to experimental study of the quantum region. Statistical rules would be of no avail if the scientist could not obtain reliable information from his experiments. The scintillation on the screen, the counter's click, must be tied to a definite causative agent. Yet according to Heisenberg's law the quantities one must have in order to do this always will escape definition. Let us see how Niels Bohr found a way through this barrier and steered atomic physics on to the path which it follows today.

When we experiment, said Bohr, what we are doing is asking a question. Behind the apparatus and instruments which we invent, behind the technical definitions which we also invent to suit our purpose, lies the question. Thus the definitions which we have used in physics reflect the questions we have asked; for "position," "velocity," "frequency" and so forth, the quantities which we summarize with p and q, are what we must measure in order to learn, How does the observed event come about? What makes it proceed as it does?

When such a question is asked, one assumes that it is possible to separate the object or process under study from the things which are done to it in the measuring experiment. This has been the method of physics and it has been a very successful method. Just because of this, one may fall into the habit of thinking that the technical definitions which accompany that method apply to every part of the physical world, that "position" always is occupied whether or not one observes this to be so. One may come to regard our system of definitions as infallible, an exact replica of the physical world, a photograph of the thing or process itself.

But now our investigation of nature has grown more detailed, our measuring techniques more refined, and we have come in the quantum region to a limitation. In any experiment it is necessary and inevitable that the measuring tool interact in some way with the subject matter. Whether our tool takes the form of matter or of light energy, it must establish some kind of contact in order to relay information to us. Unless that happens we learn nothing. And in the quantum region the smallest possible inter-action is in relation to the subject of study large and powerful. We come to recognize that the light which must be cast in order to take our photograph gravely affects the picture which is taken. We do not, as we imagined, record the precise thing or process itself. Strictly speaking, we never have. Due to the small size of Planck's constant it was possible to form the idea that our experimental intrusions always could be controlled so that we might separate the effect of the tool from the subject matter. Now we recognize that this idea does not pass the crucial test of meas-urement; we recognize that the effect produced by our experi-mental intrusion forms an inseparable part of the system we observe. The definitions of physics, the p's and q's, do not describe a system which exists apart from the effects produced in its observation.

We come to the problem which is the most fundamental of all: How are we to progress toward further knowledge of the quantum region and what lies beyond it? How are we to answer our questions? We must rely on experiment for information and this means that we must be able to connect the observed effect with a cause. Our measurements are what we know; on them we build science. We depend upon the very causal connection which Heisenberg's law denies.

Yet the same law which announces indeterminacy tells us something else and shows us how we may proceed. It tells us that the quantities we must have to make a causal connection—the inadequate q and p definitions—stand in a relationship of mutual exclusion. When one can be sharply defined the other member of the pair will not appear in the measurement situation at all. Thus we may establish the vitally important causal connection if we retain and work with partly inadequate definitions. In one sort of experimental frame what we call "position" may be tied down exactly. In a different measurement situation what we call "momentum" may, in the same sense, be precisely defined. Thus in practice we will retain our old definitions, but understand them differently. "Position" no longer will be tied to something which retains that property even when we are not looking. The same is true of "frequency" and the other definitions pertaining to particles and waves which in the past we have invented. Now they signify "an aspect which appears only under certain known experimental conditions that we impose."

Our experimental questioning thus may continue, but with a deeper understanding of the question. To a certain extent the question one asks about the quantum region will determine the nature of the reply. We must stand either on one side of the door or the other and where we stand determines which side is apparent to our view. Our understanding of the relationship between experimental method and the aspect of the quantum region which this method presents to our observation means that we may, as before, use models, the pictorial representations of the p's and q's of physics. While formerly we conceived these models as photographs of the system itself, now they become alternate understandings, depending upon the experimental viewpoint. When one picture is useful in interpreting an experiment, the other cannot be used at all. This is true of the wave model, which logically and beautifully explains the way quantity becomes quality in the world of the atom, explains why we find the specific forms of matter which exist and why they combine in the specific ways they do. For the very quantum states of energy which defy closer analysis in any single experiment account for the stability and form of what we observe directly. Now we can begin to ask why.

Under different experimental conditions we can employ the model which depicts the electron as a material particle. Our two models are not contradictory, as at first sight they appear to be, because one model fills in just what the other leaves out. They are complementary to one another; together they express the whole.

In this way Niels Bohr identified Planck's constant as the numerical recognition of a necessary indefiniteness which arises as measurements on the atomic scale are taken, measurements pertaining to definitions which do not fit nature accurately on that scale but which nevertheless must be employed. The quantum discontinuities are gaps in the physicist's knowledge of his subject matter in any single experimental situation. But they do not signify an absence of understanding, because of the complementary relationship between the definitions which is given by Heisenberg's law. Bohr's interpretation is called "complementarity," expressing the idea of a use of mutually exclusive concepts in order to reach the maximum possible understanding.

In the mathematical structure of quantum mechanics—the

precision tool of atomic physics—Bohr read a record of how physicists had expanded the old physics so that it could account for the quantum region as well as for phenomena of ordinary experience. The p's and q's of the old mechanics had been handled in a different mathematical way—for example, with a different rule of multiplication. The old wave equation had been developed so that it moved in a mathematical space of more than three dimensions. By such technical means, checks had been imposed on the older understanding of the physical world and a narrow logical framework had been broadened so that it might contain what in the old was contradictory.

Implicit in the old mechanics was the idea of a sharp distinction between the subject of observation and the tool used to observe it. Expressed in the new mechanics, which included and went beyond the old, was the recognition of a relationship between the subject and the method used in its study. Thus the solutions given by quantum mechanics were of necessity many-sided. Many different possibilities were taken into account, some being more probable than others, depending upon the experimental angle of view. The wave equation of Schrödinger, for example, did not describe the behavior of a single subatomic particle, such as the electron, but the behavior of a great number of electrons under the same conditions. With that equation one calculated the probability of being able to measure "position" or "wavelength" in a particular experiment.°

In all forms of quantum mechanics the method of statistics was built directly into the symbolism. One began with averages and probabilities; they lay behind the "certainties" of large-scale events. The deeper reality behind the old understanding of nature was realized in a statistical form of description.

A broader harmony in which the old appears in the new, a

° Max Born contributed in a major way to this interpretation of Schrödinger's symbolism as a "wave of probability."

harmony in which contradiction is resolved, this is the essence of Bohr's interpretation of quantum mechanics. It was the resolution of which he had dreamed so long, much as Albert Einstein had dreamed of finding a godlike pattern in the universe and on the basis of science appeared to find it, or at least part of it. (In the chapter which follows we will see Einstein "find" even more of a pattern "out there.") Bohr, however, did not think that his resolution of the quantum problems, or rather the resolution of others which he interpreted, reflected a logical pattern which exists apart from man. Science, including mathematics, was man's way of seeing reality, Bohr thought, man's creation. Soon we will see Einstein and Bohr in argument over this difference in their views.

After Niels Bohr returned from his ski trip to Norway, again to work with Heisenberg in Copenhagen, there was a long period during which the two men, disputing even more hotly than before, ironed out inconsistencies in their work, again through the analysis of different experiments, and (often with Pauli's assistance) brought it to a more final form. This came to be called "the Copenhagen interpretation of quantum mechanics." It is the theory which Newcomb explained to Oldfield in an earlier chapter, and it is the theory which Bohr defended in a debate with Albert Einstein.

They met in 1927 at the Solvay conference, in Brussels. Bohr had been invited to present the new Copenhagen interpretation to the physicists who gathered there every three years to discuss recent developments in their science and, needless to say, argue about them.

For many weeks Bohr worked on the paper which he would read at the conference. He had worked out his interpretation by formulating the same idea again and again, trying to see it from every possible side. The conclusions he had reached were inseparable, he felt, from the way that they had been reached. Now in his paper he tried to express this many-sided approach, struggling with what he called "inefficiencies of expression," as he had struggled in 1913 and indeed whenever he tried to express his thoughts in written form. As soon as he had completed a sentence he would see what it failed to say and begin to change it, trying to put, in that one sentence, different points of view and the connections and interactions among them. It was not perfect clarity he sought but a flexible and deep understanding. The idea could be made to seem clear but then, he felt, one's understanding of it necessarily would be shallow.

Often weeks, months, even years would go by as Bohr's struggle with sentences continued. "How can this be improved?" he would ask friends and members of his family. Gradually the first draft of a paper would darken with inserts and deletions until it barely was legible. Then the whole thing would be abandoned as another way of presenting his ideas occurred to him, often after a sleepless night and a bout of self-criticism.

"Work, finish, publish!" the Dane would exhort himself, in the words of Michael Faraday. When the paper, complete at last, went to a secretary for typing his associates would sigh with relief. But then he would recover the typed manuscript from the secretary's desk and make further changes in it. The new version, typed up, would be sent off for publication. Printers' proofs would come back to Copenhagen for checking. Bohr would seize them and begin another and sometimes major revision. Again and again Pauli would be invited to Copenhagen for his invaluable criticism until at length he wrote to Bohr: "If the last proof is sent away, then I will come." (It was when Dirac was asked for comparable assistance that he explained it was not *his* custom to begin a sentence without knowing how it was going to end.)

At times, when prolonged effort still had brought no satisfactory result, Bohr would decide to go away to a place where he could work without interruption as he had done in Norway. With a necessary assistant (a physicist who could take notes and also argue with him) he would escape to out-of-season resorts and other places which he believed would prove quiet. One of his assistants, Leon Rosenfeld, has described these quests for peace: the time, in an out-of-the-way and forlorn English hotel, when they waged "a regular war of nerves against an irascible schoolmistress for the possession of the parlour"; the time in Italy when the dog that accompanied them on their long walks attacked the local cattle and aroused the wrath of the local farmers. And the especially memorable occasion in Belgium when they settled, all unknowingly, in a gambling resort and observed a fellow, seemingly in their own line of work, who was scanning columns of figures and feverishly scribbling calculations, obviously in search of a "system." They were beholding for the first time, said Rosenfeld, "a real gambler" and they were beginning to understand why they had been greeted by the manager of their hotel with dark suspicion and been told, "Pay in advance."

Back in Copenhagen, tales of these adventures would circulate among the young men of the institute to become part of its lore. And when Bohr celebrated an important birthday, such as his fiftieth, he would be given a special birthday present, a collection of articles and poems written in his honor by students and assistants, past and present. This *Journal of Jocular Physics* was not a serious work because then, as the preface stated, Bohr might feel that it was his duty to read it "and even try to learn something." It seemed better, said the authors, to spare him "such an arduous task." And so the journal poked fun at physics and physicists, and particularly at Bohr himself. In it he could

Cover of the Journal of Jocular Physics *which was presented to Bohr on his fiftieth birthday. The Bohr-dog (inset) is an illustration for Fritz Kalckar's tale of "the mystical atom work-shop" which appears in the same issue, together with such articles as "Note on the Influence of Cosmetic Rays on the Landau-Coefficient, meaning the physicist Lev Landau." Both drawings are the work of Piet Hein.*

read a devastating parody of his writing style; see himself caricatured as a dog; and re-experience with Rosenfeld the prolonged
and painful process of completing a paper.

All of this was in fact a tribute to Niels Bohr, written in the
spirit of the man himself, who could see the comical aspect even
of matters which he took very seriously.

The people closest to Bohr were those who most appreciated
him, for in private conversation he could express himself more
fully than in an article or a lecture. He was not always in hot
pursuit of an idea as we have shown him in this book. He was
also a teacher and one who chose to teach in private conversations rather than through formal lectures. The student invited to
Bohr's study was asked penetrating questions about his research
and drawn into a lively discussion. There was no criticism. They
met to learn together, and as a result of this learning the student could criticize his own work.

A student came to know Bohr quite intimately through these
discussions of their mutual work. Rather than making a joke in
order "to break the ice" for a "serious" discussion of science,
Bohr's moods were quite evident in *all* that he said as, depending
upon his feelings that day, he played lightheartedly with mental
inventions or gloomily questioned solid ideas of the past or with
youthful enthusiasm set forth on a new track. "They come to the
scientist," Rosenfeld said, speaking of the young physicists who
were attracted to Bohr's institute, "but they find the man, in the
full sense of the word."

In the spring of 1927, Bohr packed the latest revision of his
paper on the Copenhagen interpretation of quantum mechanics
and departed for Brussels and the Solvay conference. Einstein
would be there and Bohr was eager to hear what he would say
about the new theory which in one way resembled Einstein's
own, the special theory of relativity. Both were founded on limit-

ing principles: certain concepts derived from experience in the medium-sized, low-speed domain of man's ordinary experience could not be extended to regions where speeds approached that of light or where quantities approached Planck's constant. In these far reaches of the physical world, classical physics gave way on the one hand to relativity, on the other to quantum theory.

In Bohr's eyes Einstein was the great pioneer of the second, as of the first. His own contributions were, by comparison, quite minor. It was Einstein who with his explanation of the photo-electric effect had shown the wider relevance of Planck's idea and it was Einstein, again, who had seen how to use statistical reasoning in descriptions of the atom, thus bringing this non-causal method to the forefront of physics. Now the fruit of Einstein's work was at hand. Bohr set out for the Solvay conference with, as he said, "great anticipations" to tell Einstein about it.

Before describing the debate which took place between the two men at that meeting, we first will pick up the thread of Einstein's life, which was dropped in 1913 when he returned to Germany. There Einstein completed the extension of his first theory of relativity; there he experienced a sudden change in his fortunes after that theory was given a test and responded to this change by entering the arena of political opinion; there his life began to be, as he said, "a little strange."

CHAPTER TWELVE

Albert Einstein: The General Theory

of Relativity

Just as with the man in the fairy tale everything he touched was transformed into gold, with me everything becomes newspaper noise . . .
— Albert Einstein, letter to Max Born (1920)

Wʜᴇɴ Einstein greeted Bohr at the Solvay conference in 1927, he was forty-eight years old. His hair was beginning to gray, his face to wrinkle. Fourteen years had passed since his decision to return to Germany and these years had been spent for the most part at the University of Berlin, the university of Hermann von Helmholtz and Max Planck.

From the beginning Einstein was uncomfortable in Berlin. He said that he often felt as if something were pressing on him and predicted, correctly, that the end would not be good. Berlin, it is true, provided what he wanted most, an opportunity to concentrate on his work. But he could not divorce himself entirely from the surroundings, the human climate, in which the work was done. At sixteen he had escaped from the discipline of his German school, from the mindless submission to authority which in his opinion characterized German intellectual life. Now he had accepted a position among the authorities from whom as a boy he had fled. From the beginning he stood out at the University of Berlin as someone who did not, and did not want to, belong.

219

Yet there for twenty years he remained. Although, in contrast to Niels Bohr, he preferred to work alone and would have been, he said, content to spend his days in a lighthouse, he never isolated himself from human society except in thought.

One of Einstein's biographers, Philipp Frank, has described the impression the new professor made on certain members of Berlin's academic world, specifically on Professor Stumpf, a prominent psychologist at the university on whom Einstein called one afternoon. It was customary for a new faculty member to introduce himself to his fellows by paying such a call and Professor and Frau Professor Stumpf were prepared to ask the polite questions which were the protocol on such occasions: "How do you like Berlin?" "How is your family?" ° and so forth.

The opportunity to ask them never presented itself. Einstein had called only because he had heard of Professor Stumpf's interest in space perception and thought that he might like to discuss its connections, if any, with the theory of relativity. Upon entering the Stumpf parlor, he launched at once into an explanation of his theory in relation to the problem of space, much to the embarrassment of Professor Stumpf, who was not well versed in mathematics and could not understand what Einstein was talking about. For half an hour the incomprehensible explanation continued and then—realizing suddenly that he had stayed too long, for such calls were supposed to be brief—the new professor muttered his good-bys and bolted.

Einstein liked to talk about important scientific questions but he did not like to chat. Matters of great import to those engaged in climbing the academic ladder from *Privatdozent* to full professor did not interest him. The colleague who asked Einstein,

° In 1901 Einstein had married Mileva Maritsch, a fellow student at the Zurich Polytechnic Institute. They had two sons. After Einstein returned to Berlin they were divorced and he married Elsa Einstein (a cousin), who remained his wife until her death in 1936.

"Have you heard that A has been elected to the Royal Prussian Academy of Science?" ". . . that B in his new paper has failed to cite the work of C?" might hear, in response to this offering, a loud guffaw of laughter. Not only did academic shoptalk fail to interest Einstein, it struck him as ridiculous.

So also did the meetings of that august institution, the Royal Prussian Academy of Science, the standards of which were so high that even some Berlin professors were not invited to become members. He found it hard to endure the Academy meetings, where members read their long papers, papers which often made but a minor contribution to knowledge but sounded very important, bolstered as they were by citations of every conceivable related work; where the members debated every question solemnly, sometimes heatedly, and always with absolute thoroughness, even the question of whether to publish a collection of scientific papers in one volume or two.

It was not so much the dullness of these meetings which disturbed Einstein as their scientific pretensions. He was intolerant of the everyday commerce of science, of the "easy" discoveries which help their authors to earn a reputation—and a living. Scientists should concern themselves with hard, basic questions, he thought. His attitude was idealistic in the extreme: science was "a temple," he said. Men should enter it not in order to make money, not to exercise their particular talents, but purely to serve, purely "for the sake of science herself."

As we have seen, Einstein himself was driven by a desire to understand, a desire which he called "a passion." Personal existence he found narrow, cramped, deadly. In thought he could free himself, moving away from street and city, nation and planet, to universals. Amid seemingly endless variety and multiplicity, he could discover "pre-established harmony." That the mind of man *could* recognize a fundamental order and form in the universe was for Einstein a source of endless amazement and delight. The passion ruled his life; all of his creative power was directed into a single channel. For Einstein, said a friend, "the difference between life and death . . . consisted only in the difference between being able and not being able to do physics."

Nearly everything else was, by contrast, trivial or ridiculous. For example, speaking to his neighbor in a funeral procession, Einstein confided that he could see no intrinsic meaning in such a ceremony. One attends a funeral, he said, only because of other people, the way "we polish our shoes every day just so that no one will say we are wearing dirty shoes." Again and again matters which others took seriously would call forth one of his loud, uninhibited laughs, to the puzzlement or annoyance of the person who just had confided in him and knew that what he had said was not funny.

In view of such encounters it is not surprising that some people called Einstein "childish." Not only was he uncomfortable in Berlin, he also caused many of his associates there to feel uneasy. Even his clothing proclaimed his indifference to convention. As time went on, he discarded from his wardrobe such articles as ties and belts, pajamas and socks. They were, he said, "unnecessary ballast," another wearisome aspect of personal existence; one by one they went overboard. But the simplification of dress, designed to increase his freedom, in the end may have subtracted from it. When the name "Einstein" became known outside the small circle of physicists, his appearance—not only the memorable face dominated by sad, radiant eyes, but the old leather jacket, baggy trousers, unshorn flowing hair—added to his reputation for unconventional thought, made him seem all the more unique and strange. People would visit the University of Berlin simply to look at Einstein, a tropical bird among penguins.

His great fame arrived suddenly, after a dramatic confirmation of the second—or general—theory of relativity. Here we will speak quite briefly of that theory, primarily to illustrate Einstein's individual approach to scientific problems, which bears on his later debate with Niels Bohr. Before we begin we would like to point out an essential distinction between quantum and relativity theory. Both treat the problem of tracing motion within the framework of space and time, but general relativity, a part of large-scale physics, deals with motion which because of the smallness of Planck's constant can be plotted exactly.

In working out his first theory of relativity, Einstein had been motivated by a desire to perfect physical law. Theory should display an "inner perfection," he said. Not only should it lead to fact, not only should it be based on premises which had experimental grounding, there was a third requirement: the premises of a theory should be logically simple, unforced. On that basis, he had found fault with Newton's motion laws which relied on one particular reference system, the one provided by stationary or absolute space, for judging the motion of a body. This special provision should not be necessary, he felt. It was artificial, unrelated to measurement, but more than that it was a blemish in the theory, something forced into it which made for needless complexity. He believed that it should be possible to eliminate this blemish. Laws could be found which enabled one to judge motion from any observational standpoint, any reference frame, laws which held whether one chose to say "The earth rotates once a day" or "The heavens revolve about the earth once a day," universal laws.

His first relativity theory had moved toward this goal, for he had been able to show that in the case of uniform motion the reference frame provided by absolute space was not necessary. His new motion laws did not rely on this or on any other preferred system of reference. But his work would not be done until he had found laws governing motion which deviates from a straight line and an unvarying speed: rotation, acceleration, deceleration and so forth—nonuniform motion. Soon after he sent his first theory away for publication he had begun to brood about the problem of nonuniform motion and about the forces due to inertia and gravity, which according to Newton produce that motion.

It was ten years before he solved this problem, and as in the first theory, his solution rested on notions of space and time (as well as force and mass) which differ radically from those formed on the basis of ordinary experience. This brought him into conflict with the ideas of his day. But Einstein's second theory of relativity like his first was motivated not by a wish to overthrow accepted ideas but by a compelling desire to simplify, generalize, and so perfect physical law. That was why he concentrated on the problem of nonuniform motion—and for as long as ten years—even though there was not a single experiment to point to a fault in the old structure of laws. As far as most scientists were concerned, there *was* no problem; Newton's laws accounted quite perfectly for planetary motion. There was no experiment, such as Michelson and Morley's failure to detect an ether current, to raise a doubt.

Without an experiment to point out a flaw in existing theory, how was it possible to find a new one? Where could one begin? Again it was Einstein's feeling about the form which the laws of nature should take which guided him. It led him to question a certain coincidence in Newtonian law which, all too conveniently in his opinion, made everything come out right. This was the perfect balance between the mass of a body, considered as a source of gravity in Newton's gravitation law, and the mass of a body as a measure of inertia in Newton's laws of motion.

According to the last, the amount of force necessary to change the motion of a body depends on its mass; a heavy object has greater inertia than a lighter one and therefore is harder to move *or to slow down*. To this law there was, apparently, an excep-

Albert Einstein in 1919, the year his general theory of relativity was given a successful test, the year his name began to appear in the newspapers.

tion: the case of falling. Instead of falling more slowly, a heavy object reaches the same acceleration as a lighter one. Newton accounted for this exception with his law of gravitation, which states that the attractive force exerted on one body by another increases with the mass of the attracted body. Thus, most conveniently, the force of gravitation always is exactly sufficient to overcome the inertia of a body so that all bodies fall at the same speed.

This perfect balance, this same mass constant which appeared in two different roles, looked to Einstein a bit suspicious, a bit forced. It should not be necessary to bring in a coincidence of this sort in order to explain the universe; God did not present such malicious complexities. He began to wonder whether the distinction between inertial and gravitational force might not be artificial. Could motion which was due to what Newton called "gravity" in fact be distinguished from motion due to what he called "inertia"? It could not be, Einstein found, after he had made an analysis of experiments which might be done in answer to the question. The distinction was artificial. It had no experimental support and therefore—according to the rule: "If you cannot relate it to measurement, you do not know it"—the distinction had no place in a physical theory.

This was the foothold Einstein required in order to construct a new theory, a theory which relied neither on coincidence nor on a preferred reference frame, a theory which explained the effects of gravity and inertia with a single idea, the gravitational field. Maxwell had accounted for the effects of electricity and magnetism with the idea of an electromagnetic field * which changed the properties of empty space so that magnetism, formerly conceived as a force which reached out to affect a body some distance away, now could be understood in terms of the space surrounding the attracted body. In similar fashion Einstein explained the effects of inertia and gravity as due to changes in the space surrounding the bodies affected. In his theory gravity was not a force which reached out in some mysterious way to act instantaneously on a remote body. It was rather a property of space, created by the masses distributed throughout the universe. Wherever a planet or star was present, there a gravita-

* Waves of radiation consist of electromagnetic fields.

tional field was created which changed the properties of space.

In Einstein's theory, the properties of field-distorted (or curved) space are responsible for the kinds of motion which occur. The path taken by a body is determined by the spatial terrain through which it passes, the "valleys" and "hills" which are encountered along the route. Thus according to Einstein the shortest distance, the easiest course, between two points is not a straight line but a curve around the hill-like distortions of space. Before he was able to formulate motion laws based on this idea, he had to find a geometry which described the new spatial terrain, a geometry which was not based on the idea of flat space and therefore straight lines, a non-Euclidean geometry. After a long search (during which he regretted that he had not studied more mathematics in his Zurich days) he found in Riemann's geometry the proper tool. Unlike other non-Euclidean geometries, Riemann's passed into that of Euclid for small regions of space, such as those of man's ordinary experience.

In speaking of the general theory of relativity, we have used the word "space" for what properly should be termed "space-time," a unity of the three spatial dimensions and one time dimension. To describe motion completely, the time, or when-dimension, must be used as well as the three where-dimensions. Before Einstein's first theory of relativity it had been assumed that the time dimension always was separable from the other three, as in ordinary earthbound experience. That this assumption did not hold for out-of-the-ordinary experience the first theory had demonstrated. And in the second theory, the four-dimensional blend of space and time also plays a fundamental role. Observers in different parts of the universe watching the same event would gauge that event differently; Einstein had demonstrated this. How then was it possible to know that it *was* the same event? How was it possible to distinguish among different subjective observations what was objective, common to all?

The theories of Einstein answered this question by means of the space-time framework called "a four-dimensional continuum." Observations taken in different parts of the universe must differ, but by referring them to the mathematical space-time framework, it was possible to obtain objective information. In other words, the laws of relativity hold for any reference system, any standpoint of observation. The laws are universal.

According to relativity, then, space and time are inseparable, and in the regions of the universe where matter is present the space-time unity possesses a curvature. From this it follows that the universe as a whole must possess a slight curvature; it must have a limit, a size. In this way the general theory made it possible for the first time to speculate on a scientific basis about the universe as a whole. This would be its most important consequence. But while the general theory plays an important role in today's science of cosmology, it is not as firmly established in science as Einstein's first relativity theory, for it has proved to be exceptionally difficult to verify.* True to the correspondence idea, the formulas of general theory dovetail with the formulas of Newtonian physics where these are known to hold. But as mentioned earlier, Newton's laws account quite perfectly for planetary motion. There was but one exception, as Einstein discovered when after completing his work he looked for a test of it. Since predictions from his theory diverged from Newtonian law in the case where bodies move in a strong gravitational field, Einstein checked observations of the planet Mercury, which at one point in its orbit is quite close to the sun. In this case Newton's laws do not give the observed orbit but, as Einstein found, agreement between his theory and observation is very good indeed. Still this was but a single verification. It was a second test, made just at the end of World War I, which persuaded many sci-

* There are but three experimental tests of the theory and in recent years questions have been raised about two of them.

entists that the theory had to be taken seriously. It was this verification which catapulted Einstein to world fame.

The test was made by a group of English scientists who were
able, because of a total eclipse, to photograph the stars in the
sun's immediate neighborhood. Their object was to learn
whether light from these stars was deflected by the sun's gravitational field, for Einstein's theory said that the path taken by light
was determined, like the paths of planets, by the structure of
space-time. There would be, the theory said, only a slight deflec-
tion in this case (no larger, comparatively, than a coin seen from
a distance of two miles); but the English team of scientists succeeded in detecting it.

Their findings were first announced at a meeting of the Royal
Society in London, and J. J. Thomson, president of the Society at
that time, called Einstein's work "one of the greatest achievements in the history of human thought." All of this was reported
quite lengthily in the newspapers, for the year was 1919: the war
between Germany and England just had ended and it seemed
worthy of special notice that English scientists had verified the
theory of a German. One English newspaper called Einstein "a
Swiss Jew" (technically he was a citizen of Switzerland) and,
seeing it, Einstein was amused. In Germany people now were
calling him, with pride, "a German scientist," but the English
seemed to prefer him as "a Swiss Jew." Should he ever become
unpopular, he remarked, the terms would be reversed. For the
English he would become suddenly "a German"; for the Germans, "a Swiss Jew."

His joke prophesied correctly, at least for Germany. Already in
1919 there were rumors abroad in that country that defeat in the
war was due not to military weakness, but treason. Germany, it
was whispered, had been "stabbed in the back" by pacifists and
Jews. As the years passed these whispers increased in volume.
Long before the coming of Adolf Hitler in 1933, Albert Einstein,

pacifist by conviction, Jew by inheritance, was attacked in public meetings and in certain newspapers.

Simultaneously, Einstein was, for many other Germans, a heroic figure. Before the war there had been proud boasts of Germany's military superiority and also of Germany's science. The first boast had proved hollow, but not the second. The former enemy themselves had verified and acclaimed Einstein's work, which modified that of the Englishman, Isaac Newton.

Other reasons have been offered to explain why so many Germans (together with the rest of the world) became interested in the theories of relativity. Presumably, like Einstein himself, they wanted to think about something far removed from their immediate lives, darkened and disrupted by the war, something as removed from destruction on earth as the shape and motions of the universe, the end of an absolute and inexorable time. Einstein's ideas, as interpreted by journalists, remained "news."

At the same time that the general public was trying to understand what the new theory meant, so were the physicists. For virtually all of them it was at first close to incomprehensible. Even the few who possessed the special knowledge of mathematics which was necessary in order to follow the argument had trouble with the theory. It could be comprehended only in its own terms and these were *new;* the scientist who would understand had to learn to think differently. Max Born has said that when he first began to study Einstein's theory he found it "fascinating but difficult and almost frightening." He learned to understand it only after prolonged study and discussions with Einstein himself. (Then he called it, as have many physicists since, "beautiful.") Gradually over the years the foreign ideas, through use, became more familiar but in the early 1920's few physicists were in a position to explain these ideas in nontechnical language to the general public.

This added to what already was a curious situation. It is no exaggeration to say that never before in history had so many people been interested in a theory of physics. (The quantum theory never captured such interest, although physicists considered it as significant and as startling as relativity theory.) In newspapers there were editorials about the meaning of Einstein's work. Among philosophers and religious leaders there were debates as to the connection between relativity and relativism (the idea that an ethical code, rather than being absolute, varies with human development). The theory also was called "antimaterialistic." In Russia it was attacked on that basis. Elsewhere it was attacked as "Communistic," its radical challenge to traditional scientific ideas being equated with the recent radical change in Russia's political system.

Some scientists joined in this attack—particularly, it has been said, those who were not theorists and who believed in "sound common sense." Among them was the German physicist and Nobel prize winner Philipp Lenard, whose experiments had given Einstein a basis for the photon theory of light, and who would become an early member of the Nazi party. Lenard and a few other scientists and philosophers joined a German pressure group which during the 1920's was engaged in a campaign to discredit Albert Einstein.

Germany at this time was a federal republic; the Kaiser and his Prussian army had been driven from power at the end of the war. There was unemployment, inflation and revolutionary attempts against the republic by factions at both ends of the political spectrum—for example, by the Communist sympathizers who controlled Munich for a while when Heisenberg was a youth, and also by those who wanted to reestablish the monarchy and its Prussian military arm. It was this last group who were circulating the story that Germany's defeat in the war had been caused not by a failure of the military but by the treason of Jews and pacifists. It was members of this group who (joined by some scientists and philosophers who were not necessarily politically inclined) were campaigning against Einstein, a campaign in which scientific-sounding arguments against his theories were being used as weapons in an attempt to undermine the new German republic.

Einstein was not, however, a silent scapegoat. On the contrary he had helped to draw the fire of these extremists by taking a strong stand on the very issues which inflamed them. And what Einstein said on *any* subject rapidly found its way into print and was read by the many who were curious about his ideas and about him. He was arguing for a strong world government and an end to war at a time and in a place where to be an internationalist and a pacifist was considered very close to treason, and not just by political extremists. He had begun to support the Zionist movement in its attempt to obtain Palestine as a common ground for the Jews of different countries at a time when this movement was opposed even by many German-Jewish people, especially professional people like Einstein himself.

There was no way of knowing in the 1920's that something worse was in store for the Jews of Germany than the ghetto and many people felt that the complete assimilation of Jews into German society was only a matter of time. They opposed Zionism as stressing religious and cultural differences of Jews and thus, it was felt, helping to prevent the desired assimilation. But Einstein, who did not believe in religious orthodoxy of any kind or in intense nationalism, nevertheless supported the drive to create a new Jewish nation. When he returned to Germany, he said, "I saw worthy Jews basely caricatured and the sight made my heart bleed. I saw how schools, comic papers, and innumerable other forces of the Gentile majority undermined the confidence even of the best of my fellow-Jews. . . ." A subtle dislike of one's own kind, that was as bad, he felt, as ghetto segregation. For this reason Einstein believed that "every Jew in the world should become attached to a living society to which he as an individual might rejoice to belong and which might enable him to bear the hatred and the humiliations that he has to put up with from the rest of the world."

Following the successful test of his second theory of relativity, Einstein stepped deliberately into the German political arena. He would not use his scientific theories as arguments for his social and political beliefs; he would use the fact that his every utterance was of interest to a large group of people in order to say what he felt should have a wider circulation. The fact that these utterances would make him decidedly unpopular in some circles meant little to him. What mattered was physics.

As we have seen, he regarded that as sacred. And not only did he dislike being paid to do physics, he liked even less being honored for what he had done. He said once that he was rather proud of the general theory of relativity. Someone else would have put together the special theory, he thought, if he had not done so; the time for it was ripe. That was not true of the general theory; physicists had not been interested in the problem. But praise for this or for any of his scientific achievements caused him deep embarrassment: he had done what he had for the sake of science herself.

It was "unfair and even in bad taste," he said, "to select a few . . . for boundless admiration, attributing superhuman powers of mind and character to them." The idea of him people had he called "simply grotesque"; to endure it was all but "unbearable."

After popular recognition of his work came, Einstein had still other reasons to think about the lonely existence of a lighthouse keeper. But the same honors and awe which caused him pain also gave him a heavy feeling of obligation. It would be wrong, he felt, to turn away from the people who regarded him with wonder and admiration and pursue his own pleasure, that is do physics and only physics. Einstein believed that man was bound inextricably to his fellows in an interdependence that is, he said, "our principal advantage over the beasts." Of this advantage and of the consequent obligation of each man to his fellows Einstein was keenly, or as he put it, "oppressively" aware. "A hundred times every day," he said, "I remind myself that my inner and outer life are based on the labors of other men. . . ." Using the opportunity fame brought him to circulate beliefs which he believed should be heard, which he fully knew would be attacked, Einstein lightened the heavy burden of obligation he carried. He risked his reputation, and did not mind if he lost it. The first mass meeting held by the anti-Einstein pressure group found Einstein in the audience, applauding. Praise could humiliate him; insults, apparently, could not.

By 1927 the fires set off by the English verification of the general theory of relativity were burning brightly. Everything, it seemed—not only Einstein's scientific ideas and the particular time when they appeared in the press; not only his deliberate promulgation of controversial extrascientific beliefs, but even his pungent way of expressing himself, even his appearance which proclaimed an indifference to matters which others took seriously—conspired to make him the King Midas of newspaper noise. The man who greeted Niels Bohr in Brussels at the Solvay conference was beginning to look like the well-known photographs, taken later in the United States, and had already acquired the reputation of a radical thinker, both in the realm of scientific ideas and out of it.

CHAPTER THIRTEEN

The Debate Between Niels Bohr
and Albert Einstein

In this century the professional philosophers have let the physicists get away with murder. It is a safe bet that no other group of scientists could have passed off and gained acceptance for such an extraordinary principle as complementarity, nor succeeded in elevating indeterminacy to a universal law.
  —James R. Newman, in *Scientific American*

The famous principle of indeterminacy is not as negative as it appears. It limits the applicability of classical concepts to atomic events in order to make room for new phenomena such as the wave-particle duality. The uncertainty principle has made our understanding richer, not poorer; it permits us to include atomic reality in the framework of classical concepts. To quote from Hamlet: 'There are more things in heaven and earth, Horatio, than are dreamt of in your philosophy.'
  —Victor F. Weisskopf, replying to James Newman in *Scientific American*

The relativity theories of Albert Einstein give a new meaning to ideas which lie at the root of classical physics. Yet in one important sense relativity theory is not a radical departure from traditional scientific thought. It does not challenge determinism—"the burning question," Erwin Schrödinger called it. Although in relativity space and time are redefined radically, the new space-time structure enables the physicist to obtain exact information, and as a consequence, make exact predictions. The properties of curved space-time change gradually and continuously. Motion through this frame may be plotted in causal sequence so that relativity, like classical physics, provides an understanding of how an event develops.

But in quantum theory, as we have seen, the single event may not be analyzed beyond a certain point, regardless of how space and time are defined. The effects of observation may not be separated out from the observed event. It is not possible either to understand the exact causal development or to predict the exact outcome.

This comparison of theories could not be made until 1927 when the Copenhagen interpretation of quantum mechanics was formulated. It was then, at the Solvay conference, that Bohr presented this interpretation and Einstein, to the surprise of the physicists gathered there, found it unacceptable.

For a number of years he had made no contributions to the quantum theory of atomic structure in the belief that the statistical path which others were following could not lead to a fundamental understanding. Einstein's feeling about this was not a secret. But now Heisenberg and Bohr had shown that the statistical rules rather than being a temporary resort were an expression of realities. It was Einstein's refusal to accept this which was surprising, in view of the evidence: Heisenberg and Bohr's analysis of experiments. This analysis had demonstrated that elementary particles do not fit the definitions of classical physics which make causal analysis possible. Quanta had been revealed as inevitable gaps in the scientist's knowledge of his subject matter as he views it from different angles in necessarily different experiments. On the basis of the evidence it was pointless to ask for a theory which "described the thing itself and not merely the probability of its occurrence," and yet Einstein was asking for just that. There were differences between his attitude and Erwin Schrödinger's, but both men shared the conviction that quanta someday and somehow would be eliminated, continuity and therefore determinism restored.

Niels Bohr, like the others, was surprised. He had talked with Einstein before this meeting and was aware of his attitude toward the quantum theory; but he had thought that Einstein would be guided by the rule which says that physical theory must be built upon ideas which are tied to measurable quantities. Applying this rule Einstein had banished the concept of an ether, of an absolute time interval, of a distinction between gravity and inertia. There were critics of relativity who argued that even though these concepts could not be related to measurement, they ought not to be ruled out of physics. Always Einstein had disagreed with them. Surely then Einstein would apply the same rule to the Copenhagen interpretation of quantum theory. Once he understood the study of experiments in back of that interpretation, he would ask no longer for a description of the thing itself. He would recognize that the idea of such a description had been formed on the basis of large-scale scientific experience and was meaningless where he was applying it.

But this did not happen. Einstein, instead, concentrated his great powers on an attempt to disprove the indeterminacy law, on which the Copenhagen interpretation rests, by finding an exception to it, a case where the law does not hold. He used thought experiments (experiments which in principle can be performed although technological limitations may make this impossible in practice) and on the basis of these thought experiments tried to demonstrate that q and p could be measured at once and precisely, in contradiction to Heisenberg's law. Each day Einstein presented Bohr with a new thought experiment; each day toward evening Bohr, after hard thought, was able to find a flaw in the experiment and thus refute the argument, only to be greeted the next morning by another ingenious experiment, for Einstein had used just this technique in his theories of relativity and he was a master at it. At length Paul Ehrenfest, a close friend of both men, said only half-jokingly, "Einstein, shame on you! You are beginning to sound like the critics of your own theories of relativity. Again and again your arguments have been refuted, but instead of applying your own rule that physics must be built on measurable relationships and not on preconceived notions, you continue to invent arguments based on those same preconceptions."

Undaunted by this scolding, Einstein continued. Indeed, three years later, when physicists again convened in Brussels, he greeted Bohr with a new thought experiment and this one was, Bohr said, "a serious challenge."

According to Heisenberg's law, the energy change of an event on the atomic-photon scale and the time when this event occurs can be determined no more exactly than Planck's constant. Einstein believed that he had found a case where this rule does not hold; his idea was based on the equation he had found ($E = mc^2$), from which energy may be determined if mass is known. To measure the energy of a photon, then, one could weigh it. Suppose that light were enclosed within a box, a box lined with mirrors so that the radiation would be retained there indefinitely. The box could be weighed. Then one photon could be released from it through a shutter operated by clockwork inside the box, the clock having been preset for a particular instant of time. Then the box could be weighed again. Knowing the change in mass, one could calculate, from Einstein's equation, the amount of energy which had been released. Thus the energy change in this case would be known precisely, as would the time when this event (the release of the photon) occurred.

Could Bohr find a flaw in Einstein's chain of reasoning? Could this experiment indeed measure p and q at once and precisely, in

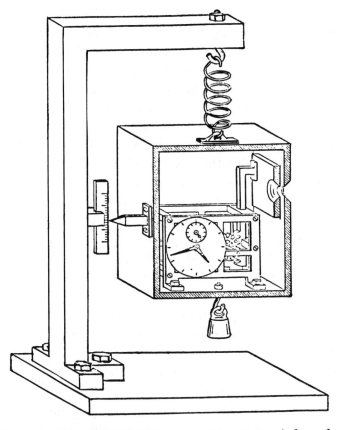

This drawing made by Niels Bohr shows how Einstein's hypothetical box containing light might be weighed by means of a rigid scale. Details have not been omitted; Bohr believed in careful analysis of the measurement process.

contradiction to the law of indeterminacy? This time when day drew to a close Einstein had not been answered. Night came; Bohr, unable to sleep, continued to wrestle with the problem. Day dawned. At last during the morning Bohr thought of something which Einstein had overlooked, an essential point: the effect of the weighing experiment on the clock.

Bohr's reasoning applied to any method of weighing but to illustrate that reasoning most clearly he chose to imagine that Einstein's box of light was hung on a spring from a rigid scale. Then when a photon was released the box would move in recoil. Its vertical position in relation to the earth's surface would change and therefore its position within the earth's gravitational field. According to the general theory of relativity, this change in spatial position would mean a change in the rate of the clock, preset and attached to the box. The change would be extremely small but in this case crucial. For due to a chain of inevitable uncertainties: the uncertainty of the escaping photon's direction, therefore of the box's recoil, therefore of its position within the earth's gravitational field, the precise time when the photon was released from the box could *not* be determined. It was indeed indeterminate to the extent given by Heisenberg's law—the cornerstone of the Copenhagen interpretation. This was the way Bohr answered the serious challenge of Einstein, who had forgotten to apply his own general theory of relativity.

When Niels Bohr returned to Copenhagen he said that he was surprised by his success, for he always had felt that he did not understand relativity theory very well. He also was keenly disappointed. Einstein, although he had granted that the Copenhagen interpretation could not be refuted on strict scientific grounds, still would not accept it as complete. The theory "yields much," he said, "but it hardly brings us closer to the secret of the Old One." Einstein still believed in the possibility of a description of the thing itself. To believe otherwise would be contrary to his scientific instinct, he said, his "inner voice."

This had guided him in the construction of his theories of relativity. As we have seen he had criticized the premises of Newtonian theory as relying on coincidence, as being needlessly complex. Einstein believed that it was just this ability to sense subtle logical blemishes, to "feel" a theory's inner perfection or the lack of it, which made it possible to recognize an underlying pattern in the universe, a pattern he called "God." That the new premises of relativity which he had found exactly represented that pattern Einstein regarded as uncertain. Just as he had shown that Newton's premises were faulty, so in the future others might show that Einstein's were inadequate. But as Einstein saw it, this would mean that the others had been able to come closer to the pattern "out there." His work in science appeared to him as a search for something outside of himself, something quite apart from himself, a search in which he was liberated from a "merely personal" existence, and in his own words he found that existence "narrow . . . deadly." The search could have an end, the liberation was possible, because man *could* comprehend. Because he possessed this instinct for logical perfection, this sense of formal, mathematical beauty, he might move closer to the Old One's secret.

Albert Einstein's inner voice spoke in this fashion. But Niels Bohr also possessed an inner voice. Like Einstein's, it had helped to guide his scientific work; and Bohr's voice spoke differently. It said that while man seeks to understand a reality which exists outside himself, the explanations which he finds for this reality are worded in human language, wholly man-made. As languages of words are invented by man, said the voice, so is the language of symbols, mathematics; it is more refined, less cumbersome than the other languages, but like them it reflects man's way of thought and not a pattern external to himself.

In Niels Bohr's view man was central: man who was in life a spectator of nature, including human behavior, but also a part of nature, an actor. Corresponding to these very different roles were different understandings and different approaches. As a spectator, said Bohr, man tries to understand the causes of human behavior on a rational basis. He tries also to judge that behavior, pardoning or condemning it, in the attempt to be impartial. But as an actor he is not guided by pure reason or by pure impartiality. Lifted out of context, he said, one sees a contradiction between these different approaches. Judged in the abstract, is not *strict* justice contradictory to *perfect* compassion? But his inner voice would reply, There is no real contradiction here, for man is the inventor of what he calls "the abstract." As spectator, he erects the framework through which "contradiction" becomes apparent.

Niels Bohr, rather than seeing a pattern external to man, recognized a great range and richness of human experience, calling for a wide variety of different approaches, science among them. The mathematical logic which accounts for the real world was not less wonderful for being the expression and creation of man. Albert Einstein in that same logic saw a way to reach a wonder apart from man.

The two physicists had discussed their differing views, or philosophies, during the Solvay conference of 1930. Following Bohr's success with Einstein's box experiment they had gone for a walk together. And Einstein on that occasion had said that he was greatly disturbed because Bohr could not see what to him was so evident.

Einstein said, in effect: "It is no use trying to explain anything with physics if you have no basis for judging that explanation in terms of its inner perfection." To work without such a basis would be, he said, "a betrayal of physics."

Bohr had replied that a physical explanation should be tied to measurable quantities, agree with all the observational facts, and its logic should not be self-contradictory. He recognized no other principle, no other guide pertaining to a theory's inner beauty. In seeking to understand an entirely new realm of nature, said Bohr, such a guide could not be trusted. To work on the assumption that one's accustomed principles would continue to hold in the new realm also—*that*, said Bohr, would be "a betrayal of physics."

Thus the new quantum theory had passed all the tests except the one which Einstein recognized and Bohr did not. Insofar as Einstein's objections had taken the form of assertions which could be tested, they had been answered, and Einstein had granted that on these grounds the theory was unassailable. Insofar as his objections were philosophical, however, they could neither be tested nor settled. Nor could they be relinquished: Einstein did not become deaf to his inner voice. He continued to argue that the theory was incomplete; the discontinuities someday would be eliminated. He continued to find fault with the Copenhagen interpretation; and Bohr continued to answer Einstein's criticism. Increasingly, theirs became a debate between two philosophers rather than two scientists; they were arguing the theory in terms of its contribution to general knowledge. In replying to Einstein, Bohr refined the Copenhagen interpretation, a problem which occupied him more and more in the years after 1930. In his thoughts, he said, he always was arguing with Einstein.

In physics, meanwhile, quantum mechanics and the Copenhagen interpretation of it remained the only theory to account fundamentally for the behavior of atoms and systems of atoms. It is today the major tool of atomic physics; by and large, scientists do not expect it to be replaced by a theory which describes "the thing itself and not merely the probability of its occurrence."

There are, however, a few exceptions. The debate between Albert Einstein and Niels Bohr, although they no longer are alive, continues, as indicated by the two quotations which introduce this chapter. In recent years the debate has taken on added interest because physicists anticipate a new theory of the structure of matter.

Broad as is the range of quantum mechanics, physicists, experimenting with higher and higher energies, appear to have found a limitation. As the poem on page 88 indicates, the quantum theory "holds in the shell as well as the core" of the atom.

But under conditions of extremely high energy, the properties which characterize the atom disappear. When in today's giant accelerators the nucleus is bombarded with subatomic particles having energies of millions and billions of electron volts, new forces and particles come into play. They pose questions which it has not been possible to answer on the basis of quantum mechanics. A new theory is needed. Judging from past experience, physicists believe that such a theory will represent a radical change in their thinking. Will the new ideas which are introduced make it possible, after all, to eliminate quanta and describe the thing itself? Or will the new theory leave the indeterminacy principle and the Copenhagen interpretation essentially unchanged? No one knows. It is not possible to say in advance what might, or might not, be accomplished on the basis of a new idea.

The next and concluding chapter of this book concerns a few of the things which happened to the men of whom we have written here after the year 1930, when Bohr answered Einstein's serious challenge.

CHAPTER FOURTEEN

Afterward

. . . convey to wider circles . . .
—Niels Bohr

ALL THE PROTAGONISTS of this book, from Rutherford and Planck to Paul Dirac, lived for a long time after 1930, when Bohr answered Einstein's objections to the quantum theory (and all in different years won the Nobel prize for their contributions to that theory). Some are living today and are active in physics.

Max Born is retired now from the University of Edinburgh, Scotland, where he taught for many years after leaving Göttingen. He lives in West Germany. Louis de Broglie, former Sorbonne professor, lives in France where he always has lived. Werner Heisenberg and Paul Dirac are working in nuclear physics, the first at the Kaiser Wilhelm Institute for Physics, now relocated in Munich and renamed the Max Planck Institute; the second at Cambridge University. They are among the physicists, most of them younger men, who seek a new general theory of matter, a theory to account for what is known about the nucleus in the way that quantum theory accounts for the data of chemistry, a theory which might eliminate the indeterminacy principle, and might not.

Ernest Rutherford, who discovered the nucleus, was the first to knock protons out of it by artificial means and thus change one element into another. His experiments, which ushered in the science of nuclear physics, were done at the end of World War

I—and except for one laboratory helper, Rutherford was working alone. The apparatus he used in these experiments resembled the one used to detect the nucleus: alpha-particle bullets from a radioactive source, target and scintillation screen. The whole setup was light enough to be picked up easily, small enough to be set on a card table and to a great extent handmade. This was soon to change. Rapidly drawing to a close was the era of laboratory glassblowers, of setups held together with wire and sealing wax, of tools for atomic research which could be carried literally under a hat—as one physicist when crossing borders carried the radioactive element he used in his experiments, so that he would not have to wait while customs officials attempted to classify the unfamiliar article.

At the end of the war, Rutherford left Manchester to direct the Cavendish Laboratory, where once he had been a student. His good friend J. J. Thomson had become master of one of the Cambridge colleges and Rutherford took over J.J.'s former position. At the Cavendish the New Zealander continued his nuclear experiments and directed the research of a new team of "boys," boys who learned that the best way to obtain more knowledge of the nucleus was to use great quantities of high-energy atomic particles as bullets. To speed up their projectiles they developed more and more powerful electromagnetic fields, bigger and bigger machines to produce those fields.

". . . We, in the Cavendish, turn out the real solid facts of nature," Rutherford bragged. He was intensely proud of his boys, one of whom, James Chadwick, discovered the neutron; and until his death in 1937 Rutherford took an active and noisy part in their work, the "fight with the machines," one of them called it. His craving to find out more did not slacken with the years nor did his ferocity; in private the boys referred to their leader as "the crocodile" because, as one explained, "The crocodile cannot turn its head. . . . It must always go forward with all-devouring jaws."

The accelerating machines which "grew" at the Cavendish also sprouted up elsewhere during the 1930's; at Berkeley, California, for example, where the cyclotron was invented. More and more money was required for the research equipment of physics. Increasingly this money came from government, particularly after research of the 1930's led to nuclear fission and to the recognition that the immense energies produced could be used as a weapon. At the turn of the century Prince Frederick of Germany had gone to a physics professor, Hermann von Helmholtz, for military advice. After 1939, the same thing was done on a far larger scale by an increasing number of governments.

At the same time that the equipment of physics was growing so was the number of men (and women) who were attracted to that science. It is possible to measure in terms of publications and man-power the growth of world-science during the last three centuries and this growth has been exceptionally rapid. Every ten or fifteen years the size of science has doubled. This means, the physicist-historian Derek J. de Solla Price said recently, that "80 to 90 percent of all the scientists that have ever lived are alive now."

During the 1930's and 1940's the center of the rapidly expanding science of physics shifted from Europe to the United States and to a lesser extent, Great Britain. Once the common language of physicists was German; today it is English.

The movement from Europe began in the early days of Hitler. It has been estimated that between 1933 and 1938 the Nazis exiled nearly two thousand top-flight scientists because of their so-called "non-Aryan" ancestry or their political beliefs. Erwin Schrödinger left Berlin in 1933, the year he and Paul Dirac shared the Nobel prize in physics. During the war Schrödinger taught at Oxford University and later at the Dublin Institute for Advanced Studies. He died in Vienna in 1961.

Max Born was among the seven professors who were "retired," suddenly, from the University of Göttingen a month after Hitler gained power. James Franck of the Franck-Hertz experiments left Göttingen soon afterward for Copenhagen and then the United States.

Albert Einstein chanced to be visiting the United States when Hitler became Chancellor of Germany; he decided not to return. The Nazis replied by confiscating his personal property and burning his works on relativity. The private campaign against him had become official. Rather than waiting for the Royal Prussian Academy to expel him, Einstein by letter offered his resignation. He did this, it is said, out of consideration for his friend Max Planck, long an officer of the Academy and its most prominent member. Planck had recognized Einstein's work very early, had brought him to Berlin, had defended him against the attacks of other German scientists. It was inevitable, Einstein thought, that the Academy would move to expel him; he wished to spare Planck the pain of this act.

After 1933, Einstein worked at the privately endowed Institute for Advanced Study in Princeton, New Jersey. Like the other institute members, he was not required to teach or perform other academic duties. He could concentrate on his own research. It was the same sort of arrangement which years before had drawn him back to Germany.

During the war Paul Dirac and Wolfgang Pauli also were associated with the Institute for Advanced Study at Princeton, but except for the war years Pauli (who won the 1945 Nobel prize for his exclusion principle) taught at the Polytechnic Institute in Zurich. He came often to the United States in the summer for physics conferences held on various campuses here and his expenses for these trips were paid. Universities had learned that if Pauli was present, other eminent physicists were attracted to the spot. "Get Pauli," a saying went, "then the others come free."

As before, they came for criticism. The young physicists, like the older ones, tried to emulate the perfection of form which characterized Pauli's work and when they completed something asked themselves, "What will Pauli say to this?"

The Viennese theorist made a number of contributions to nuclear physics. Like Heisenberg and Dirac he was trying hard to find a new general theory of matter. Once, during the 1950's, it looked as if Pauli, in collaboration with Heisenberg, might be close to getting it. In New York, Pauli presented their ideas to an audience of physicists which included Niels Bohr, who was making one of his quite frequent visits to the United States. When Pauli finished talking the criticism began. Many in the audience found fault with the new theory, particularly some of the younger men. When the discussion was over, Bohr summed it up in terms of the lesson physicists had learned from relativity and quantum theory about common sense. "We are all agreed," said Bohr, "that your theory is crazy. The question which divides us is whether it is crazy enough to have a chance of being correct."

As it turned out, the theory was discarded, like many other fruitless attempts. Pauli died in 1958, when the mysteries of the nucleus, which for twenty years he had been trying to understand, still were unsolved. He never found out.

Albert Einstein was another who died (in 1955) without knowing the outcome of work to which he had devoted himself for several decades. The photon theory of 1905 had been inspired by Einstein's wish to unify in some way the essentially discontinuous atomic theory of matter and the continuous wave (or field) theory of radiation. As we have noted, his work in quantum theory continued to be inspired by this wish. And with the completion of his general theory of relativity, Einstein began to look for an elaboration of the four-dimensional space-time continuum, an elaboration which would make it possible to account on the basis of continuity for the observed discontinuities of elementary particles and photons. He completed several versions of his unified field theory, but no way ever has been found to verify it. The theory did not offer a prediction which both differed from accepted theory and could be tested.

Like Einstein, Niels Bohr lived to be more than seventy-five years old. In 1932 he was given a castle which the brewer of Carlsberg Beer had donated for the lifetime use of Denmark's most outstanding intellectual. In the castle on the brewery grounds the Bohrs housed and entertained the physicists who always were visiting Copenhagen. "It was a grand residence," said one with awe. "There were many rooms with blackboards." Another, who received an invitation to dine at the castle, felt that his wardrobe was inadequate for the occasion and appeared, to everyone's surprise, in a splendid uniform. It was red in color, "screaming" red. It was in fact the uniform which is worn by the postmen of Denmark, from one of whom it had been borrowed.

Many of the physicists who visited Copenhagen during the 1930's did not return to their homes. They had come in response to letters from Bohr suggesting that perhaps they might like to visit him for a while and consider a permanent change of residence, in view of the threatening political situation in Europe. One physicist, who left Italy in 1938, for political reasons, had not been able to bring much money with him. He had been doing cosmic-ray research and soon after he arrived in Copenhagen, Bohr, largely for his guest's benefit, began to organize a conference on this subject, inviting specialists from many countries to attend. After the conference was over, the Italian physicist was told that Bohr would like to see him in his office. There the Dane talked for quite a while, but more unintelligibly than usual and in an exceptionally low voice. He appeared to be embarrassed. All the listener could catch was something about ". . . and see the secretary to get the check." When he obeyed these instructions he discovered that Bohr was paying him for attending the conference which Bohr had arranged for him.

A great number of scientists received Bohr's embarrassed assistance; many, due to his influence, were able to leave Europe for positions in England and the United States. But not everyone felt able to accept the invitations to Copenhagen.

When Max Planck received one he told a friend that he could not accept. "On my previous travels," he said, "I felt myself to be a representative of German science and was proud of it. Now I would have to hide my face in shame."

As well as banishing scientists the Nazis were trying to rub out their ideas. Nazism, said the Minister of Education, "is not the enemy of science, but only of theories." Like relativity, the quantum theory was labeled "Jewish physics" and banned at the universities.

Werner Heisenberg was one of the German scientists who openly opposed this policy. Planck, who then was in his seventies, remained silent. He was, said friends, unable to change from an obedient servant of the state to an active opponent, a rebel. Rather than anger, he felt shame and a sense of duty, a duty to stand by his country and try to save what he called "German science" from the Nazis.

Once, early in the Nazi regime, Planck, a brave man in his own way, attempted to argue with Adolf Hitler. As director of the Kaiser Wilhelm Institute, it was Planck's duty to make an annual call on the head of the German government and he used this opportunity to speak to Hitler about a great chemist who, because he was Jewish, was being forced out of Germany. Hitler would not listen to Planck's arguments. He said that the Jews were Communists, that nothing could divert him from his "great goal." "Do not think that I have such weak nerves. . . . Everything will be carried out to the last letter," he shouted at his audience, one old gentleman.

After the Nazis' policies changed from the exile of Jews to their extermination, Planck must have known of it, for his son Erwin was part of an anti-Nazi movement and it has been said that he kept his father well informed.

Once there had been four Planck children. Their mother had died in 1909, followed by the deaths during the years of the First World War of three children. Only Erwin was left.

Planck had married again and fathered another child. His research, his activities in various scientific societies, his mountain climbing had continued until he was an old man; and he had survived the intensive bombings of Germany at the end of the Second World War, although he lost his home and once was buried for several hours when an air-raid shelter, hit by a bomb, caved in. Friends said that his will to live remained strong until he learned the fate of his son.

Erwin had been among the conspirators who, toward the end of the war, attempted to kill Hitler. When the carefully planted time bomb missed its target, Erwin like the others was captured and given a terrible death by the Gestapo.

When he learned of this Planck was silent. He sat down at the piano and began to play. Later he wrote to a friend, "You give me credit for too much if you think I have the strength to withstand this pain."

Planck died at the age of eighty-nine in Göttingen. He had been taken there during the war in a military car, sent by Americans when they learned that his refuge near the Elbe River had been destroyed and that he was stranded between the advancing Allied armies and the retreating Germans.

Werner Heisenberg was another who refused an invitation to

Copenhagen. Like Planck, he believed that it was his duty to remain in his country and try to protect "German physics" from the Nazis. During the war he became head of the German scientific project which was working toward construction of a uranium pile; the attempt to develop atomic bombs, using the material which can be produced in such a pile, never began. But until the Allied invasion of Germany the progress of this uranium project was unknown. Physicists outside Germany did know, however, that such a project existed and knowing it, feared the possible outcome—the Nazis armed with an atomic bomb.

Prompted by this fear, a group of physicists who had come to the United States from Europe asked Albert Einstein to write a letter to President Roosevelt explaining that atomic energy could be used for destruction and warning him that German scientists almost certainly were working on this. The same physicists, joined by many others, had gone to work on the atomic-bomb project which had been set up as a result of their explanations and warnings.

Niels Bohr joined them in 1943. Denmark was occupied by the Germans at that time and when they began large-scale arrests of "non-Aryans" Bohr, whose mother was Jewish, was in danger. The Danish underground helped him to escape, as they helped many others. He went first to Sweden, in a small fishing boat; then was flown to England. After his departure members of the German S.S. came to the Copenhagen institute and searched Bohr's files (while his secretary watched, telling them to keep everything in good order). Presumably they were seeking scientific secrets which had military application; but what they found were letters to Bohr from his friend—Professor Heisenberg—of Germany. The war had separated the two men; it had strained but not ended their friendship.

This was not an exceptional case. For example, the physicist Sam Goudsmit who toward the end of the war was sent by the

United States government to Germany on a mission to investigate that country's wartime scientific work—and to arrest its authors—did not regard each and every one of these German scientists as his personal enemy—even though Goudsmit's Dutch parents were among those who "disappeared" forever in a Nazi concentration camp. Seeing Heisenberg for the first time after his arrest by other members of the American mission, Goudsmit said, "I greeted my old friend and colleague cordially."

After Niels Bohr came to England he was told of the progress of the atomic-bomb program in the United States. Robert Oppenheimer, former director of the Los Alamos, New Mexico, branch of that program, has said that from the beginning the English were much involved in it, "much more than is generally known in this country." When Bohr heard that atomic bombs were on the way to becoming a reality, he began to think about what the existence of such weapons would mean in the future after the war was over. That Russia would learn quite soon to make atomic bombs was, he thought, inevitable; no keeping of scientific secrets could prevent it from happening. If the scientists of one country could get the bomb so could those of others with a sufficiently advanced technology.

Bohr knew scientists in Russia and was acquainted with the harsh political conditions in that country and the suspicion of the West which did not bode well for the future. Unlike some other optimists he did not think that Russia and the Western nations then allied in war would find it easy to maintain friendly relations after the war was over. There would, he thought, be tension between the different economic and political systems, an extremely dangerous tension once Russia and the United States both were armed with atomic bombs. Bohr believed therefore that planning with Russia for control of the new weapons should begin at once, before post-war tensions developed, before the weapons were a reality, having, in the context of East-West tension, the appearance of a threat. Such a course would mean international inspection systems; it would mean giving Russia knowledge of atomic energy. He thought, however, that this gift also would profit the giver because Russia, drawn for her own benefit into a situation of dropped barriers and information freely exchanged, could not remain the same country as before. In other words, East-West tension could pose a problem in the future; and to quote Robert Oppenheimer, Bohr "wanted to change the framework in which this problem would appear early enough so that the problem itself would be altered."

In England, speaking along these lines, Bohr tried to persuade Prime Minister Churchill and his advisors of the importance of immediate overtures to Russia. With the same object in mind he came in 1943 to the United States. There he would lend his technical services to the Los Alamos bomb project. But that was not his main purpose in coming to this country.

The story of how, four years earlier, Bohr brought to the United States the first news of the discovery of atomic fission (experiments done in Germany and interpreted in Sweden) has appeared in print many times. Not quite so well known is the story of how he tried to avert what he felt might be the consequences of that discovery, of his memorandums and visits to the political and military advisors of President Roosevelt, and after his death, President Truman; of his open letters to the new United Nations.

After the bomb was made and the war ended, Bohr's work—like many another physicist's—remained political as well as scientific. His proposals had not been adopted. He had come to think that international control of atomic (and when they came, thermonuclear) weapons could not become a reality until the distrust between East and West diminished. So that different peoples might have the fullest opportunity to learn about each other—a necessary condition, he believed, for the growth of confidence—he advocated the elimination of national barriers to travel and to the exchange of ideas and information. In moving toward a more open world he thought that science would play a leading role, for in science as in no other human endeavor, national considerations were out of place, obsolete. The scientists formed an international family.

Until his death in 1962, Bohr worked in different scientific enterprises to strengthen the bonds of this family. And he remained director of the Copenhagen Institute for Theoretical Physics. Now he has been succeeded by one of his sons, Aage Bohr, who is also a physicist.

It seems fitting to end this final chapter with a last glimpse of our main protagonists, Niels Bohr and Albert Einstein, and in 1948 both men chanced to be working in the same spot, the Princeton Institute for Advanced Study. Indeed, Bohr was using Einstein's office at that time. (Einstein did not like the big room he had been given and had moved into one adjoining it, which was smaller and had been intended for his assistant.) In Einstein's office Bohr began to work on an article about his debate with Einstein.

Twenty years had passed since they first had matched wits on the question of quantum mechanics, but the debate still went on. A few years before he moved into Einstein's office Bohr once again had been trying to win over his unyielding adversary, with the usual outcome. After that long argument, the Dane had sought out a close friend and told him bitterly, despairingly, "I am sick of myself."

But now, in Einstein's office, Bohr again was retracing the old arguments with Einstein, again trying to improve the formulation of his ideas. By describing the actual occasions when he had debated with Einstein, Bohr wanted, he said, to show "how much I owe to him for inspiration" and also to "convey to wider

This photograph of the blackboard in Bohr's working room at Carls-berg castle shows his rough sketch of the Einstein box of light. He drew it the evening before his death when during a conversation he was explaining and developing his ideas.

circles an impression of how essential the open-minded exchange of ideas has been. . . ."

He was hard at work on the article when Abraham Pais, another physicist member of the institute, stopped in to see him. Of course Bohr was not actually writing: his head down, his brows knit, he was, said Pais, "furiously pacing around the table in the center of the room."

Would Pais please help him out? Bohr asked. Would he take down a few sentences as they emerged from his thought?

Pais agreed and sat down at the table, paper and pencil ready, while Bohr continued to orbit the table, occasionally uttering a few words in his soft voice. (It has been said that in order to take dictation from Bohr it was necessary "to execute a continuous rotation at the same rate as the orbital motion.")

The dictation came slowly; often Bohr would repeat the same word over and over again. "He would dwell on one word," said Pais, "coax it, implore it, to find the continuation." And on this day one of the repeated words chanced to be "Einstein." In deep thought and almost running around the table, Bohr was repeating, "Einstein . . . Einstein."

The continuation he sought did not come and walking over to the window, he gazed out, still coaxing the same thought, "Einstein . . . Einstein."

Just at that moment Pais saw the door to the office slowly begin to open. It opened very softly—Bohr did not hear it—and someone tiptoed into the room: Einstein.

"Then," said Pais, Einstein "beckoned to me, with his finger on his lips, to be very quiet, his urchin smile on his face." Silently, he made a beeline for the table at which Pais was sitting.

And just as he arrived there, Bohr, still oblivious of the visitor, still muttering "Einstein" at the window, suddenly appeared to have gotten hold of the elusive words, and uttering a firm "Einstein" he turned around.

"There they were," said Pais, "face to face."

Like the others, Bohr was struck dumb by this uncanny manifestation of his thought, this specter he had summoned forth, this irrepressible Einstein. Speechless, the three men stared at each other.

Then Einstein explained why he had come—for tobacco. Smoking was against his doctor's orders, but the doctor, he explained, had forbidden him to buy tobacco; he had not forbidden him to steal it—in this case from the tobacco pot which rested on the table in Bohr's office. With the explanation, said Pais, "the spell was broken.

"Soon we were all bursting with laughter."

FURTHER READING

Biographies, Autobiographies, and Reminiscences

Andrade, E. N. da C. *Rutherford and the Nature of the Atom.* New York: Doubleday and Company, Inc., 1964.

Birks, J. B., ed. *Rutherford at Manchester.* New York: W. A. Benjamin Inc., 1963.

Born, Max. *My Life: Recollections of Nobel Laureate.* London: Taylor and Francis, 1978.

Dirac, P. A. M. *The Development of Quantum Theory.* New York: Gordon and Breach, 1977.

———. "Recollections of an Exciting Era." In *History of Twentieth Century Physics,* edited by C. Weiner. New York: Academic Press, 1977.

Einstein, Albert. "Autobiographical Notes." Vol. 1, *Albert Einstein: Philosopher-Scientist,* edited by Paul Arthur Schilpp. New York: Harper and Brothers, 1959.

Heilbron, J. L. *The Dilemmas of the Upright Man: Max Planck as Spokesman for German Science.* Berkeley: University of California Press, 1986.

Heisenberg, W. *Physics and Beyond: Encounters and Conversations.* New York: Harper, 1971.

Moore, R. *Niels Bohr.* New York: Knopf, 1966.

Pais, Abraham. *'Subtle is the Lord . . .' The Science and Life of Albert Einstein.* New York: Oxford University Press, 1982.

Planck, Max. *Scientific Autobiography and Other Papers.* New York: Philosophical Library, 1949.

Pupin, Michael. *From Immigrant to Inventor.* New York: Charles Scribner's Sons, 1960.

Thomson, J. J. *Recollections and Reflections.* London: Bell, 1936.

Wiener, Norbert. *Ex-Prodigy: My Childhood and Youth.* New York: Simon and Schuster, 1953.

Wilson, David. *Rutherford: Simple Genius.* Cambridge: The M.I.T. Press, 1983.

T. Kuhn and his associates have amassed a vast and extremely valuable *Archives for the History of Quantum Physics* which includes interviews with many of the leading contributors, as well as notebooks, letters, and other materials.

See also the entries for Bohr, Born, Pauli, etc., in C. C. Gillispie, ed., *Dictionary of Scientific Biography* (Charles Scribner's Sons, 1973).

Introductions to Modern Physics

Amaldi, Ginestra. *The Nature of Matter.* Chicago: University of Chicago Press, 1986.

Andrade, E. N. da C. *An Approach to Modern Physics.* New York: Doubleday and Company, Inc., 1957.

Barnett, Lincoln. *The Universe and Dr. Einstein.* Rev. ed. New York: The New American Library, 1952.

Born, Max. *The Restless Universe.* New York: Dover Publications, 1951.

Bunge, M., and W. R. Shea, eds. *Rutherford and Physics at the Turn of the Century.* New York: Dawson and Science History Publications, 1979.

Einstein Albert, and Leopold Infeld. *The Evolution of Physics.* New York: Simon and Schuster, 1961.

Feinberg, Gerald. *What is the World Made Of?* New York: Doubleday, 1978.

Fierz, M., and V. S. Weisskopf, eds. *Theoretical Physics in the Twentieth Century.* New York: Interscience, 1960.

Hendry, John. *The Creation of Quantum Mechanics and the Bohr-Pauli Dialogue.* Dordrech: D. Reidel, 1984.

Hoffmann, Banesh. *The Strange Story of the Quantum.* New York: Dover Publications, 1959.

Hund, F. *The History of Quantum Theory.* New York: Harper and Row, 1974.

Jammer, Max. *The Conceptual Development of Quantum Mechanics.* New York: Wiley, 1966.

Kuhn, T. S. *Black Body Theory and Quantum Discontinuity 1894–1912.* Oxford: Claredon Press, 1978; Chicago: University of Chicago Press, 1987.

Mehra, J., and Rechenberg, H. *The Historical Development of Quantum Theory.* Vols. 1-5 (further volumes in preparation). New York: Springer, 1982–.

Pagels, Heinz. *The Cosmic Code.* New York: Simon and Schuster, 1982.

Pais, Abraham. *Inward Bound.* New York: Oxford University Press, 1986.

Russell, Bertrand. *The ABC of Relativity.* Rev. ed. Fairlawn, N.J.: Essential Books, 1958.

Segre, Emilio. *From X-rays to Quarks: Modern Physicists and their Discoveries.* New York: W. H. Freeman, 1980.

Trenn, T. J. *The Self-Splitting Atom.* London: Taylor and Francis, 1977.

Weinberg, Steven. *The Discovery of Subatomic Particles.* New York: W. H. Freeman, 1983.

Weisskopf, Victor F. *Knowledge and Wonder.* New York: Doubleday and Company, Inc., 1963.

Woolf, Harry, ed. *Some Strangeness in the Proportions: A Centennial Symposium to Celebrate the Achievements of Albert Einstein.* Reading, Penn.: Addison-Wesley Publishing Co., 1980.

Modern Physics Considered in Relation to Other Knowledge

Bohr, Niels. "Discussions with Einstein." In *Albert Einstein: Philosopher-Scientist*, edited by Paul Arthur Schilpp. Evanston, Ill.: Harper and Brothers, 1949.

————. *Atomic Theory and the Description of Nature*. London: Cambridge University Press, 1961.

Einstein, Albert. *Ideas and Opinions*. New York: Crown Publishers, 1954.

Fine, Arthur. *The Shaky Game: Einstein's Realism and the Quantum Theory*. Chicago: University of Chicago Press, 1986.

Forman, Paul. "Weimar Culture, Causality and Quantum Theory, 1918-1927." *Historical Studies Physical Sciences* 3 (1971): 1-116.

Hacking, Ian. *Intervening and Representing*. New York: Cambridge University Press, 1983.

Heisenberg, Werner. *Physics and Philosophy: The Revolution in Modern Science*. New York: Harper and Brothers, 1962.

Kuhn, Thomas S. *The Structure of Scientific Revolutions*. Chicago: University of Chicago Press, 1962.

MacKinnon, Ernan. *Scientific Explanation and Atomic Physics*. Chicago: University of Chicago Press, 1982.

Schilpp, Paul Arthur, ed. *Albert Einstein: Philosopher-Scientist*. Vol. 1. New York: Harper and Brothers, 1959. Included here is Bohr's account of his debates with Einstein.

Scott, W. T. *Erwin Schrodinger: An Introduction to His Writings*. Amherst: University of Massachusetts Press, 1966.

On Physicists and Public Affairs

Dickson, David. *The New Politics of Science*. New York: Patheon, 1984.

Goudsmit, Samuel A. *Alsos*. New York: Henry Schuman, 1947.

Hewlett, R. G., and O. E. Anderson. *The New World 1939/1946*. State Park: Pennsylvania State University Press, 1962.

Jungk, Robert. *Brighter than a Thousand Suns: A Personal History of the Atomic Scientists*. New York: Harcourt, Brace and Company: 1958.

Kevles, Daniel J. *The Physicists*. New York: Vintage, 1979.

Price, Derek J. de Solla. "Diseases of Science." Chapter 5 in *Science since Babylon*. New Haven: Yale University Press, 1962.

Sherwin, Martin J. *A World Destroyed*. New York: Knopf, 1975.

Smith, A. K. *A Peril and a Hope*. Chicago: University of Chicago Press, 1965.

York, H. F. *The Advisors*. New York: W. H. Freeman, 1976.

Index

Index